The Guernsey Cattle Breeders' Yearbook for 1899

General Information on Guernsey Cattle, Rules and Registration

by American Guernsey Cattle Club

with an introduction by Jackson Chambers

This work contains material that was originally published in 1899.

This publication is within the Public Domain.

This edition is reprinted for educational purposes
and in accordance with all applicable Federal Laws.

Introduction Copyright 2017 by Jackson Chambers

Self Reliance Books

Get more historic titles on animal and stock breeding, gardening and old fashioned skills by visiting us at:

http://selfreliancebooks.blogspot.com/

Introduction

I am pleased to present another title in the "Cattle" series.

The work is in the Public Domain and is re-printed here in accordance with Federal Laws.

As with all reprinted books of this age that are intended to perfectly reproduce the original edition, considerable pains and effort had to be undertaken to correct fading and sometimes outright damage to existing proofs of this title. At times, this task is quite monumental, requiring an almost total "rebuilding" of some pages from digital proofs of multiple copies. Despite this, imperfections still sometimes exist in the final proof and may detract from the visual appearance of the text.

I hope you enjoy reading this book as much as I enjoyed making it available to readers again.

Jackson Chambers

The American Guernsey Cattle Club.

President, JAMES M. CODMAN, Brookline, Mass.

Vice Presidents, { LEVI P. MORTON, Rhinecliff, N. Y.,
{ SYDNEY FISHER, Knowlton, P. Q., Canada.

Secretary and Treasurer, WM. H. CALDWELL, Peterboro, N. H.

Executive Committee, JAMES M. CODMAN, FRANCIS SHAW, N. K. FAIRBANK, JAMES LOGAN FISHER, EZRA MICHENER, HENRY PALMER, E. N. HOWELL, CHAS. L. HILL, WM. H. CALDWELL.

Members, January, 1899.

Andrews, J. H.,	Farmington, Ct.	Evans, Joseph,	Marlton, N. J.
Axtell, L. V.,	Perry, Lake Co., Ohio.	Eustis, W. E. C.,	Readville, Mass.
Apgar, Allen S.,	New York City.	Fairbank, N. K.,	Chicago, Ill.
Bahnson, Henry T.,	Salem, N. C.	Farwell, John V.,	Chicago, Ill.
Barbour, H. W.,	Farmington, Ct.	Fay, H. H.,	Wood's Holl, Mass.
Beach, C. M.,	Hartford, Ct.	Ferrell, C. B.,	Columbus, Ohio.
Beirne, Jas. H.,	Oakfield, Wis.	Fisher, James Logan,	Philadelphia, Pa.
Biddle, Craig,	Andalusia, Pa.	Fisher, S. A.,	Knowlton, Canada.
Borden, R. A.,	Easton, N. Y.	Fox, J. M.,	Foxburg, Pa.
Bowditch, N. I.,	Framingham, Mass.	Foote, M. H.,	Spring Prairie, Wis.
Bradley, S. R.,	Nyack, N. Y.	Forsyth, James,	Owego, N. Y.
Brunson, Jas. L.,	Langhorne, Pa.	Freeman, E. C.,	Cornwall, Pa.
Buckley, P. B.,	Valley Falls, N. Y.	French, J. D. W.,	Boston, Mass.
Bullene, I.,	Lawrence, Kansas.	Fuller, A. F.,	Catasauqua, Pa.
Bush, J. J.,	Elmira, N. Y.	Fuller, J. W.,	Catasauqua, Pa.
Cabot, Louis,	Brookline, Mass.	Gill, E. T.,	Haddonfield, N. J.
Caldwell, Wm. H.,	Peterboro, N. H.	Graeff, James McK.,	Westport, N. Y.
Cameron, J. D.,	Harrisburg, Pa.	Graeff, Frances A.,	Westport, N. Y.
Cameron, Colin,	Lochiel, Arizona.	Greenshields, J. N.,	Montreal, Canada.
Campbell, George,	Philadelphia, Pa.	Griscom, C. A.,	Haverford P. O., Pa.
Canson, Le G. B.,	Burlington, Vt.	Hadwen, O. B.,	Worcester, Mass.
Carr, A. M.,	Salem, Ohio.	Haines, Z.,	West Grove, Pa.
Carter, Charles S.,	Lenape, Pa.	Harvey, W. B.,	West Grove, Pa.
Cassatt, A. J.,	Berwyn, Pa.	Hately, Walter C.,	Chicago, Ill.
Chalmers, John C.,	Ann Arbor, Mich.	Hearts, Benj.,	Charlottetown, P. E. I.
Christie, P. H.,	Clove, N. Y.	Heisey, S. C.,	Elizabethtown, Pa.
Clark, Edw. S.,	Cooperstown, N. Y.	Hicks, H. D.,	Old Westbury, N. Y.
Codman, James M.,	Brookline, Mass.	Higgins, J. C.,	Delaware City, Del.
Coleman, Robert H.,	Cornwall, Pa.	Hill, Chas. L.,	Rosendale, Wis.
Cutting, Walter,	Pittsfield, Mass.	Hill, Geo. C.,	Rosendale, Wis.
Davison, G. H.,	Millbrook, N. Y.	Hoard, W. D.,	Fort Atkinson, Wis.
Deming, C.,	Farmington, Ct.	Hope, Joseph I.,	Madison, N. J.
Duke, James B.,	Somerville, N. J.	Houston, S. F.,	Philadelphia, Pa.
Duncan, Jos. M.,	Silver Spring, N. J.	Howell, E. N.,	Poughkeepsie, N. Y.

Howland, A. M., Trustee,	Dona, Ana P. O., N. M.
Hoxie, Sam'l L.,	Leonardsville, N. Y.
Hughes, Mark,	West Grove, Pa.
Hunter, J. H.,	Valley Falls, N. Y.
Jones, J. Arthur,	West Hopkinton, N. H.
King, Chas. R.,	Andalusia, Pa.
King, John A.,	Great Neck, N. Y.
Kinsell, D. R.,	Columbus, O.
La Monte, Geo.,	Bound Brook, N. J.
Lawrence, James,	Groton, Mass.
Lewis, C. W.,	Farmington, Ct.
Livingston, Johnston,	Tivoli, N. Y.
Livingston, J. H.,	Tivoli, N. Y.
Logan, A. Sydney,	Philadelphia, Pa.
Lord, Robt W.,	Boston, Mass.
McBurney, Dr Chas,	New York City.
Merriam, Herbert,	Weston, Mass.
Michener, Ezra,	Carversville, Pa.
Mitchell, Ehram B.,	Harrisburg, Pa.
Mixter, Geo,	Boston, Mass.
Moore, James,	Philadelphia, Pa.
Morgan, J. Pierpont	New York City.
Morrell, Edward,	Torresdale, Pa
Morton, Levi P.,	Rhinecliff, N. Y.
Norton, Paul T.,	Somerville, N. J.
Page, R. H., Jr,	Columbus, N. J.
Palmer, Henry,	Avondale, Pa.
Palmer, J. H.,	Jewett City, Ct.
Paul, W. M,	Moorestown, N J.
Peck, Corydon,	Locke, N. Y.
Pierce, Henry,	San Francisco, Cal.
Porter, Thomas E.,	Andover, Ct.
Porter, Miss S.,	Farmington, Ct.
Redmond, G.,	Tivoli, N. Y.
Robbins, S. W.,	Wethersfield, Ct.
Roberts, Percival Jr.,	Philadelphia, Pa.
Rogers, Jacob C.,	Peabody, Mass.
Scott, I. R.,	Ward, Pa.
Sharpless, Thomas,	West Chester, Pa.
Shaw, Francis,	Wayland, Mass.
Smith, C. Morton,	Philadelphia, Pa.
Smith, J. Gregory,	St. Albans, Vt.
Solveson, Chas.,	Nashatah, Wis.
Spencer, S. S.,	Lancaster, Pa.
Stuyvesant, R.,	New York City.
Taylor, Clayton C.,	Lawton's, N. Y.
Taylor, H. A. C.,	Newport, R. I.
Taylor, R. J.,	Owego, N. Y.
Thomas, Geo. C.,	Philadelphia, Pa.
Thompson, C. J.,	Farmington, Ct
Tillotson, E. W.,	Farmington, Ct
Tomlinson, W. I.,	Marlton, N. J.
Tratt, F W.,	Whitewater, wis.
Treadwell, G H,	Albany, N. Y.
Tuttle, Howard B.,	Naugatuck, Ct
Twombly, H. McK.,	Madison, N. J.
Walker, P. G.,	Cecil, Pa.
Warner, Alexander,	Baxter Springs, Kan.
Warner, Benj. S.,	Baxter Springs, Kan.
Watts, G. S.,	Baltimore, Md.
Whitcomb, L. F.,	Florida, Mass.
Wilhelm, Isabel S.,	Harrisburg, Pa.
Willets, S. P. T.,	Roslyn, L. I., N. Y
Young, B. L.,	Auburndale, Mass.

Sheet Anchor, No. 2934, A. G. C. C.
A Magnificent Show Bull and Sire.
Owned by H. McK Twombly, Madison, N. J.

Guernsey Interests.

From a careful study of the Herd Register for the *Guernsey*, we find that the interest in the breed has largely increased within the last few years. The time has come when the Guernsey takes her place among dairy cattle, second to none in the production of the richest, golden colored milk, cream or butter. She has also proven herself in all trials where equal conditions have been given her, as the most economical butter producer. She is fast proving herself as a cow that can be depended upon to make good yearly records for butter. A Guernsey can reasonably be expected to give during the year 9,000 to 12,000 lbs. of milk, or from 350 to 700 lbs. of butter.

Best dairy experience and knowledge has shown that a *flush test* for a *short period* is a *very poor guide* for the *profitableness of a dairy cow*. It is not the flashy show yard trial, where fresh cows heavily fed appeal to a sportive crowd. Neither is it the seven day test, which is always made when the cows are at their best, that we *most* desire. It is the answer to the question always asked by all intelligent students of dairying, "Will the cow pay me to summer and winter her, and may I hope she will perpetuate in her offspring those habits of persistent and profitable dairy production? In this connection we do not wish to be understood as antagonistic to seven day butter tests. They are of great interest and value. We feel, however, that it is the "day after day" work that indicates the real value of an animal, and is of more far-reaching importance.

For twenty years the Guernsey has won friends wherever introduced by her honest work in private dairies. Within the last few years she has publicly spoken of these modest endeavors in such a way as to awaken interest and command respect. In this issue is recorded a compilation of the best milk and butter records of Guernsey cows. It is the largest, we believe, that has ever been attempted. A great deal of care has been taken in securing the facts given, yet we do not believe the lists are complete. There are undoubtedly a great many animals that should appear in them. We trust corrections or additions will be forwarded the Secretary's office that they may appear in some reprint in the future.

Whatever growth has been made in the work of the Club during the past few years, in no instance does that of any year equal that for the year just closed. By studying the records of the last six years we find many interesting comparisons. Taken as a whole the entries and transfers for the year ending Dec. 1, 1898, is nearly double the number for five or six years ago. Then when we compare the number of entries for each year with that of the preceding one, we find that the most marked increase is between the last two years, or those ending in December, 1897 and 1898.

Entries for the Last Seven Years.

ENTRIES	1892 Bulls	1892 Cows	1893 Bulls	1893 Cows	1894 Bulls	1894 Cows	1895 Bulls	1895 Cows	1896 Bulls	1896 Cows	1897 Bulls	1897 Cows	1898 Bulls	1898 Cows
December	23	25	19	57	27	47	16	33	35	71	31	105	42	102
January	37	66	22	41	26	61	22	52	26	52	30	65	54	118
February	16	49	31	51	34	37	28	45	24	31	33	65	45	74
FIRST QUARTER	76	134	72	149	87	145	66	130	85	154	94	239	131	294
March	38	63	43	67	26	69	35	94	49	139	46	79	58	97
April	46	81	41	86	23	45	30	44	44	98	26	41	42	54
May	10	38	40	61	15	44	37	89	34	54	39	73	46	63
SECOND QUARTER	100	182	124	214	64	158	102	227	127	291	111	102	146	214
June	12	38	26	29	15	25	27	62	30	63	32	54	47	57
July	17	23	37	29	28	46	40	97	23	53	27	61	43	45
August	34	66	36	51	38	53	37	77	48	78	30	81	55	111
THIRD QUARTER	63	167	99	109	81	124	104	205	101	194	89	186	145	263
September	30	62	29	72	39	69	37	58	33	62	56	66	41	120
October	20	46	48	49	27	64	36	43	24	28	71	96	50	90
November	35	53	17	37	36	94	37	80	26	53	36	46	32	42
FOURTH YEAR	85	161	94	158	102	227	110	181	83	148	157	208	123	152
YEAR	324	604	389	630	334	654	382	743	396	787	451	835	545	1023

Transfers for Last Seven Years.

TRANSFERS	1892	1893	1894	1895	1896	1897	1898
December	43	88	47	79	62	126	196
January	59	53	34	62	90	98	102
February	69	63	48	71	77	75	88
FIRST QUARTER	171	204	129	212	229	299	416
March	54	73	57	107	80	89	145
April	62	141	81	182	102	76	113
May	68	75	45	147	85	119	99
SECOND QUARTER	184	289	183	436	267	284	357
June	29	63	30	132	79	65	90
July	36	86	51	91	61	168	85
August	51	33	48	49	26	79	104
THIRD QUARTER	116	182	129	272	166	312	276
September	81	40	61	65	35	92	94
October	118	81	76	72	68	94	93
November	91	63	74	115	81	112	61
FOURTH QUARTER	290	184	211	252	184	298	248
YEAR	761	859	652	1172	846	1193	1300

While the transfers do not show so marked an increase while comparing the work of 1897 with that of 1898 as do the entries, yet the same growth is apparent when the work of the five or six years is studied.

This increase seems largely attributable to a general awakening of interest in the breed and a realization of the true merits of the Guernsey; to a desire on the part of progressive dairymen for a cow having size and constitution with a square, open, business-like conformation; a cow that will give a good flow of milk that has a high per cent. of butter fat and is of a rich, natural color; one that when a dollar is invested in food will return in the hands of any careful manager the greatest profit. In all this the Guernsey has proven herself as being in the front rank.

These figures give encouragement to all Guernsey breeders, for not only do they show a greater demand for the cattle, but a careful study also discloses the fact that there has been a large number of sales of bulls to parties who wish to grade up or improve their dairy herds. This puts the breed in contact with new persons, and paves the way for the introduction of registered herds. The advent of the glass bottle and sale of milk graded upon its value as food; the prominence given to the necessity of healthy animals; and to the rapid advance in dairy knowledge attributed to the agricultural press and to the many dairy schools, have caused the dairymen to select and improve their herds that they may hold their place in advancing competition. All this had a lasting influence to strengthen the *Guernsey* as *the ideal business dairy cow*.

Financial Statement, A. C. G. C., Year Ending Dec. 1, 1898.

Wm. H. Caldwell, Treasurer.

Receipts.		Expenditures.	
From Entries,	$2,952 00	For Salary,	$1,400 00
Transfers,	1,306 00	Office Help,	720 00
Membership Fees,	400 00	Office Supplies,	543 60
Private Herd Books,	59 50	Postage,	268 00
Herd Register,	44 25	Herd Register,	1,226 05
Magazine (sub.)	237 35	Private Herd Books,	166 16
Magazine (adv.)	195 00	Travelling and Meetings,	355 00
Sketches and Pedigrees,	5 75	Circular Matter,	210 30
General Account,	293 30	General Account,	317 92
Total Receipts,	$5,493 15	Total Expenditures,	$5,237 93
Balance on hand December 1, 1897,	801 97	Balance in Bank, December 1, 1898,	1,057 19
	$6,295 12		$6,295 12

Imp. Lord Stranford 2187 A. G. C. C.
One of the best known Guernsey Bulls in America. Owned by Mr. Jas B Duke.
Somerville, N. J.

The History of the Guernsey.

During the last few months there has been numerous requests for information regarding the *Guernsey*. Several asked if there was not some book giving their origin, development, introduction to this country and special merits. This has been one purpose of these quarterly issues of the Register.

It has been said that to know a breed of cattle you must be acquainted with their nature, their nurture and their ideal. That is, you must realize the foundation from whence they came, what their surroundings have given them, and what results they are accomplishing.

Several volumes of the Herd Register have been completed. In Volume I. we find an interesting statement regarding the formation of the Club and the establishment of the Register. Both Mr. James M. Codman, now President, and Mr. Silas Betts, the late President of the Club, give detailed accounts of the condition of the cattle on the Island of Guernsey, and their introduction to this country. Volumes II., III., and IV., are all wholly devoted to the registrations and transfers as placed on the books of the Club. In Volume V. is is given a portrait of Mr. Edward Norton, the much respected Secretary, whose earnest endeavors were largely instrumental in establishing the Club and its Register upon a good foundation. A sketch reviewing his work accompanies his likeness.

With Volume VI., January, 1895, came a change in the publication of the Register. For over three years it has appeared in quarterly parts, in each of which several pages are devoted in recording matters of passing interest in the progress of the breed in this country and on the Island. A single issue may not seem of much interest. As one looks back, however, through the various numbers and find animals of note illustrated and described ; portraits and sketches of those who have been prominent among the breeders of the Guernsey ; illustrations and descriptions of various stables and herds ; milk and butter records as brought forward, as well as much other matter relating to conditions in this country, in England and on the Island, he is impressed that all these will be of great value in the years to come.

It is by this means that a history of the breed is being written, and to this must be referred such inquiries as are constantly being received at the office.

The individual breeder has an important part to play in this work. These pages will be of more interest and the history of greater value, inasmuch as the matter recorded covers more fully all facts relating to the breed. These should be forwarded to the Secretary's office that they may be used. Never was more interest shown in the breed nor more business done through the Club's office. What is now needed is the concerted action of all breeders to gain for the Guernsey the position she deserves in the fore front of dairy cattle.

THE HERD REGISTER.

Volume I. includes Parts 1, published December, 1878; Part 2, May, 1881; Part 3, April, 1883; and Part 4, January, 1884. The volume was issued in 1884, and contains entries of Bulls Nos. 1 to 338 and of Cows Nos. 1 to 1394.

Volume II. was published in 1887, and includes Parts 5, 6, 7 and 8, and contains entries of Bulls Nos. 540 to 1212 and Cows Nos. 1395 to 2826.

Volume III. was published in 1891, and includes Parts 9, 10, 11 and 12, and contains entries of Bulls Nos. 1213 to 2220 and Cows Nos. 2827 to 4393.

Volume IV. was published in 1892, and contains Parts 13, 14, 15 and 16, and contains entries of Bulls Nos. 2221 to 2861 and Cows Nos. 4394 to 5784.

Volume V. was published in 1894 and includes Parts 17, 18, 19 and 20, and contains entries of Bulls Nos. 2862 to 3572 and Cows Nos. 5785 to 7080.

Copies of the first five volumes may be had from the Secretary's office, bound in paper covers, for $2.00 per volume; in half leather for $2.75.

The above five volumes of the Herd Register were issued previous to January, 1895. Since that time the Herd Register has appeared in quarterly parts of from 60 to 100 pages each, and been known as

THE HERD REGISTER AND BREEDERS' JOURNAL.

A Quarterly Magazine of 60 to 120 Pages. Issued January, April, July and October. The only publication devoted wholly to the interests of the Guernsey. Gives general information, tests, as well as entries and transfers of Guernseys, and is edited from the office of the Club, thus being the *official organ* of the Guernsey interests. It takes the place of the Herd Register as previously and irregularly issued.

Aside from enabling the publication of the record of entries and transfers more promptly, each issue devotes several pages to matters of interest to all breeders of Guernseys. Special pains are taken to make the publication a magazine of the highest class, well illustrated, and of increasing value to all interested in the Guernsey.

This is published upon the subscription basis. Subscriptions are payable in advance, and the magazine is stopped at the expiration of the subscription.

VOLUME VI.

Part 21, issued Jan. 1895,	Bulls Nos. 3573 to 3667	Cows Nos. 7081 to 7239
Part 22, " April "	" " 3668 to 3759	" " 7240 to 7392
Part 23, " July "	" " 3760 to 3851	" " 7392 to 7540
Part 24, " Oct. "	" " 3852 to 3942	" " 7541 to 7678
Part 25, " Jan. 1896,	" " 3943 to 4064	" " 7679 to 7852
Part 26, " April "	" " 4065 to 4245	" " 7853 to 8181
Part 27, " July "	" " 4246 to 4439	" " 8182 to 8562
Part 28, " Oct. "	" " 4440 to 4522	" " 8563 to 8962

This double volume can be had, bound in a manner similar to the first five volumes, at the following prices:

Complete issues, including Journal portion, bound in paper covers, $2.00; in half leather, $4.75. Register portion only, in paper, $3.00; in half leather, $3.75.

VOLUME VII.
   Part 29, issued Jan. 1897. Bulls Nos. 4523 to 4633   Cows Nos. 8963 to 9157
   Part 30,   "   Apr.   "   "   "   4634 to 4716   "   "   9158 to 9351
   Part 31,   "   July   "   "   "   4717 to 4826   "   "   9352 to 9593
   Part 32,   "   Oct.   "   "   "   4827 to 4937   "   "   9594 to 9732

   This volume can be had bound in a similar manner to the preceding ones, at the following prices:
   Complete issues, including Journal portion, bound in paper covers, $2.00; in half leather, $2.75. Register portion only, in paper, $1.50; in half leather, $2.25.

VOLUME VIII.
   Part 33, issued Jan. 1898. Bulls Nos. 4938 to 5045   Cows Nos. 9733 to 9752
   Part 34,   "   Apr.   "   "   "   5046 to 5182   "   "   9953 to 10172
   Part 35,   "   July   "   "   "   5183 to 5318   "   "   10173 to 10463
   Part 36,   "   Oct.   "   "   "   5319 to 5461   "   "   10464 to 10657

   This volume can be had bound in a similar manner to the preceding ones, at the following prices:
   Complete issues, including Journal portion, bound in paper covers, $2.00; in half leather, $2.25. Register portion only, in paper, $1.50; in half leather, $2.25.

   Volume IX. commenced with Part 37, January, 1899.

   Members are allowed $1 reduction on each of the above volumes, and on subscriptions to the Register and Journal.

Subscriptions per year, $2.00. Members' Subscription, $1.00.

   A limited space is devoted to advertisements. The rates for same are given on application.

GRAND COMBINATION OFFERS:

The Herd Register and Breeders' Journal with **Hoard's Dairyman,**
One Year for Only **$2.25**

   Hoard's Dairyman, published weekly and edited by Ex-Governor Hoard at Fort Atkinson, Wis., is the recognized authority on Dairy Interests.

The Herd Register and Breeders' Journal with **Country Gentleman,**
One Year for Only **$3.25**

   The Cultivator and Country Gentleman is published by L. Tucker & Son at Albany, N. Y., and has long been classed in the fore front of Agricultural weeklies.

The Herd Register and Breeders' Journal with **Rural New Yorker,**
One Year for Only **$2.50**

   The weekly Rural New Yorker has a well known reputation among Agriculturists and Horticulturists. Published in New York City.

The Herd Register and Breeders' Journal with **National Stockman and Farmer,**
One Year for Only **$2.45**

   The National Stockman and Farmer is devoted to Agriculture, Stock Husbandry and Home Interests. Its specialties are the Markets and Business Side of Farming. It is published weekly at Pittsburg, Pa., Buffalo, N. Y., and Chicago, Ill.

The Herd Register and Breeders' Journal with **Western Agriculturist and Live Stock Journal,** One Year for Only **$2 25**

A semi-monthly devoted to the science of breeding, feeding and management of Stock, with special departments for different classes of stock and the Dairy. Published by T. Butterworth, Chicago, Ill.

The Herd Register and Breeders' Journal with **The Ohio Practical Farmer,** One Year for Only **$2.40**

The great National Agricultural, Live Stock and Home Journal. Published weekly at Cleveland, Ohio.

THE ISLAND OF GUERNSEY HERD BOOKS.

The office will supply these to breeders. There are 9 volumes of the Royal and 9 parts of the General Herd Books. These can be had at $1.00 per volume or part.

THE ENGLISH GUERNSEY HERD BOOK.

There are 14 volumes of same which can be secured through the office.

Group of Guernseys.
Owned by Mr. P. H. Christie, Clove, N. Y.

Butter Test for Guernsey Cattle.

It is twenty years since the first organized effort on the part of the admirers of the *Guernsey* to bring her before the public. During this time the recognition the breed has had has been very encouraging. It is a pleasure to note how universally is the tendency to measure the value of a Guernsey by her actual work in the dairy. How much milk or how much butter the animal gives, if she is well built, with perfect udder formation, is her choicest recommendation. Breeders should be encouraged in their efforts in this direction. The Guernsey Breeders' Association, composed of leading breeders around Philadelphia, have done much to create an interest in this work. The results of their tests brought forward three five-hundred pound cows. Such work very forcibly came to the attention of the Executive Committee of the American Guernsey Cattle Club, and to encourage breeders to keep records of the work of their animals, the Club has offered $300 in prizes.

With the great advance of dairy instruction and the care shown by dairymen in eliminating from their herds all unprofitable animals, it is of great importance to each breeder, and to the interest of the breed in general, to know as much of the capabilities of each animal as is possible. It is to encourage this work that these inducements are offered, and it is hoped many breeders will avail themselves of them.

It is not remarkable yields for a short time that enhances the value of an animal. It is the ability to consume and make profitable return for the food given them. In this the *Guernsey* has proven *par excellence* by the honest trials made of the breeds by the Agricultural Experiment Stations, and in the exhibition at the World's Fair in 1893.

What of the future? Will not such honest efforts reap their reward? They cannot help but so doing. As we look back after the next ten or twenty years, we will find the Guernsey *the business cow*. There will be developed from the foundation now being laid, an animal in size, disposition, strength of constitution, formation of udder, and the converting of food into dairy products most economically, that will appeal to all as the **" Ideal Business Cow."**

The American Guernsey Cattle Club offer the following premiums for the cows or herds of Guernseys making the three best records for butter-fat for one year, under the conditions hereafter named :

For Individual Cows, $50, $30, $20.
For Herds of Five Cows, each, $100, $60, $40.

Conditions of Test.

1. All animals competing must be registered in the Herd Register of the American Cattle Club.

2. The following entry fees shall be paid to the Treasurer of the Club on receipt of notification from him that the animals named will be accepted for competition, but no animal or animals shall be enrolled unless said fee is paid before the opening of the test. For each cow, $5. For each herd entered, $15.

3. Each contestant shall be allowed to name seven animals for the herd prize, the results to be determined from the records of the five best animals.

4. These tests shall be under the supervision of the Executive Committee of the American Guernsey Cattle Club, and any member of the Executive Committee owning animals competing in said tests shall be barred from having any supervison of said test or tests. All cows shall be wholly under the control of the owner so far as feeding and general treatment are concerned.

5. All the expenses connected with the tests shall be paid by the contestants except those incurred by carrying out the provisions contained in Rule 9.

6. Each year's test shall commence November 1st; the first test commencing Nov. 1, 1898. All animals competing shall be named at least 30 days prior to the opening of each test. The results of each test shall be reported to the annual meeting of the Club, when the prizes will be awarded.

7. At the end of each month every contestant shall report to the office of the Club upon blanks furnished them for such purpose by said office:

a—A complete record of the weights of each milking.

b—An approximate statement of the amount and kind of food given the animals, and as to the manner of stabling and care of same, including dates of service or when in heat or not served.

8. About the middle of each month samples shall be taken of the night's and the following morning's milk and sent to the Agricultural Experiment Station of the state in which the animal is located, or to such place as may be directed or approved by the Executive Committee, these samples to be properly labeled with the date and amount of each milking. The result of such tests to be reported by the tester to the office of the Club.

9. At such times as the Executive Committee supervising said test or tests shall see fit, but at least twice during the year, they shall send anyone whom they may deputize, to visit the herds from which animals are entered, to weigh and test the milk from cows competing.

10. The results of each year's tests shall be computed in the following manner: The weights of milk produced each month shall be multiplied by the per cent. of butter-fat as shown by the official test for that month, and the sum of the results thus obtained shall be the year's record.

The Entries in the Competition for 1898 are:

FOR HERD PRIZE.

Geo. C. Hill & Son, Rosendale, Wis., with—

Madame Tricksey 6519,	6 yrs. old.	Calved Apr. 14, '98
Lady Bishop 6518,	7 "	" Mar. 29, '98
Benjamin's Primrose 7820,	4 "	" Dec. 29, '98
May Bishop 8604,	3 "	" Nov. 13, '98
Nounon 6569,	9 "	" Mar. 14, '98
Prestoun 6570,	7 "	" Nov. 26, '98
Countess Bishop 1868,	4 "	" Mar. 20, '98

Ezra Michener, Carversville, Pa., with—

Jennie H. 4348,	6 yrs. old.	Calved Sept. 11, '98
Maid of Orange 4155,	9 "	" Oct. 23, '98
King's Myra 5339,	8 "	" Oct. 23, '98
Queen Bee 6168,	7 "	" June 13, '98
Lady Thorne 8503,	4 "	" Sept. 14, '98
Miranda 7th 3465,	11 "	"
Mary Marshall 3d 7371,	4 "	" Aug. 9, '98

Levi P. Morton, Rhinecliff, N. Y., with—

Imp. Doutta Galla 4th 7675,	7 yrs. old.	Calved Apr. 15, '98
Imp. Louise of Ellerslie 8930,	3 "	" Oct. 10, '98
Imp. Villet's Gem 8664,	5 "	" Mar. 16, '98
Quibble 8017,	6 "	" Oct. 25, '98
Buda 7178,	6 "	" Mar. 5, '98
Decasse 7669,	4 "	" Jan. 24, '98

FOR SINGLE COW PRIZE.

Jas. H. Beirne, Oakfield, Wis., with—

Lily Ella 7240,	5 yrs. old.	Calved Dec. 7, '98
Lilyita 7241,	5 "	" Dec. 7, '98

Geo. C. Hill & Son, Rosendale, Wis., with—

Countess Bishop 7869,	4 yrs. old.	Calved Mar. 20, '98

Levi P. Morton, Rhinecliff, N. Y., with—

Imp. Doutta Galla 4th 7675,	7 yrs. old.	Calved Apr. 15, '98
Imp. Louise of Ellerslie 8930,	3 "	" Oct. 10, '98

Fill Pail des Ruettes 2d 3864, A. G. C. C.
A noted prize winner. Owned by Jas. Forsyth, Riverside Stock Farm, Owego, N. Y.

Valencia, A. G. C. C.

SOME NOTED GUERNSEY COWS
WITH
BEST MILK AND BUTTER RECORDS.

Until recently, Guernsey in America were kept chiefly for family use. Guernsey breeders have *never attempted an organized effort* to establish records for their cattle. The results obtained during the past twenty years with the Guernsey have come from straightforward everyday work in the dairy. We find none of what are usually termed *forced* tests. In connection herewith appears a résumé of what has been collated from the past. In every instance it is reasonable to expect that the work may be duplicated; furthermore, I believe as good and possibly better work will yet be found to have already been done, but is not included, as never having been reported to the office or publicly.

This list is given as a basis, with the hope that before the issue of another Year Book it may revive many additions, and any errors may be corrected.

Records covering a long period and giving so fully the details are of great value, and Guernsey breeders should be encouraged in making them. They present a marked contrast from the excessively high records for a short time which until recently have been held as a measure of a dairy animal.

Best Milk Record.

Lily Alexandre, 1059, holds the best year's milk record. She was dropped on the Island of Guernsey in August, 1879, and imported to this country in 1882. When nine years old she made the following record:

October, 1888,	1306	pounds milk.
November, "	1218½	" "
December, "	1132½	" "
January, 1889.	1010¾	" "
February "	912	" "
March, "	1067½	" "
April, "	1047½	" "
May, "	1236½	" "
June, "	1132½	" "
July, "	1016	" "
August, "	953¾	" "
September, "	822½	" "
Total for one year,	12856	" "

There was but one test made of this, and that well along in the period of lactation. The test showed 7.2 per cent. butter fat.

Lily Alexandre 1059 A. G. C. C.

Lily Alexandre was owned by Mr. Mark Hughes, of West Grove, Pa., at the time of her record, and is now the property of Mr. Francis Shaw, Wayland, Mass.

Year's Milk Records Over 10,000 Lbs.

NAME AND NUMBER.	OWNER.	YEAR.	LBS.
Lily Alexandre 1059	Francis Shaw,	1879	12,856
Miss Bobolink 2157	N. I. Bowditch,	1897	12,437
Imported Bretonne 3660	Levi P. Morton,	1894	11,218¾
Quibble 6017	Levi P. Morton,	1898	10,548
Vanessa 2100	Sydney Fisher,	1893	10,504
Passagere 2d 1582	Levi P. Morton,	1896	10,316¼
Imported Esmeralda 8657	Levi P. Morton,	1898	10,174
Lady Beth 3926	J. R. Huston,	1896	10,031
Cousin Beatrice 7303	Levi P. Morton,	1898	10,023
Coralie 446	William M. Paul,	1888	10,000

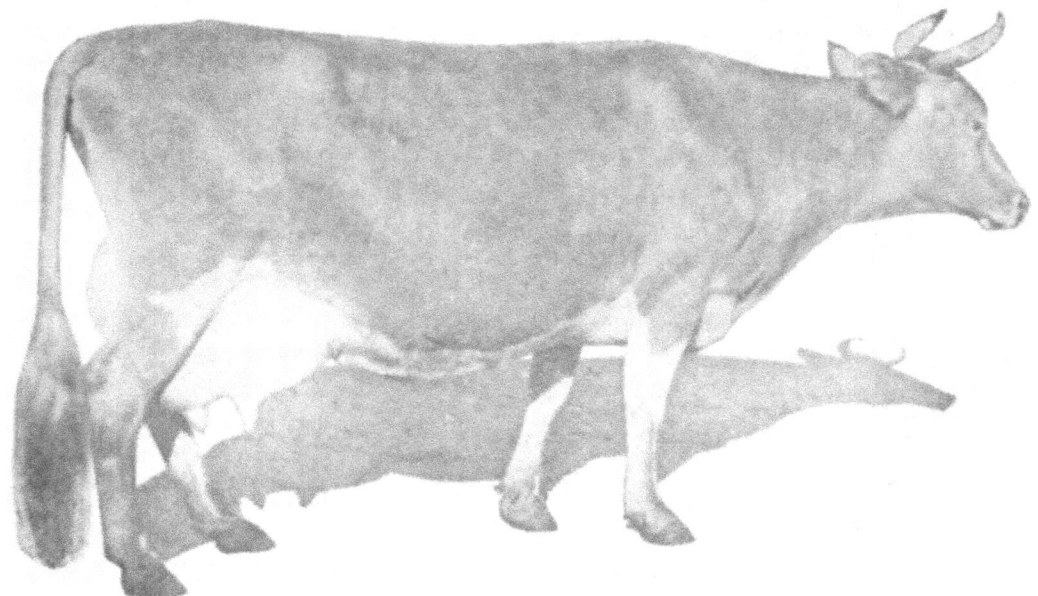

Pretty Dairymaid 2d 6366, A. G. C. C.

Owned by Francis Shaw, Wayland, Mass.

For a shorter period, Mr. Shaw's imported Pretty Dairymaid 2d of Guernsey 6366, gave an official test on the Island of Guernsey of:

    61 lbs. 2 oz. milk first day.
    62 lbs. 12 oz. milk second day.
    52 lbs. 9 oz. milk third day.
    ―――――――
    176 lbs. 7 oz. milk in three days.

BEST BUTTER RECORD.

Imported Bretonne, 3660, owned by Hon. Levi P. Morton, of Rhinecliff, N. Y., holds the best butter record for a Guernsey in one year. Bretonne was dropped on the Island of Guernsey in August, 1884, and was imported in November, 1887. When ten years old she completed a year's test, as is shown by the following carefully kept account of her work and management during the year ending October 20, 1894. He stable was a comfortable box stall in the finely appointed barn at Rhinecliff, and she always went out with the rest of the herd for exercise in the yards. During the hot days of summer she was kept in the barn daytimes and turned in the barnyard nights. The same good care and usual amount of food allotted to the best producing portion of the herd was given her. Although well fed, as will be seen in the records that follows, she was not forced to eat beyond what she showed a ready inclination for. Careful records were kept. The grain was weighed every day, and the ensilage and hay occasionally. Her milk was carefully weighed each milking. The butter fat determinations were made by the Babcock tester the middle of each month, each test being of a composite sample of eight consecutive milkings. The result was that from Oct. 20, 1893, to Oct. 20, 1894, Bretonne gave 11,218¾ pounds milk. Highest test, 6.1 per cent., lowest test, 5.2 per cent. butter fat, or 602.91 pounds butter fat, equivalent to 703.4 pounds butter, ⅙ added to fat.

In order to study the work more in detail, the following data is given:

MILK AND BUTTER FAT RECORD.

1893.				
October (20-31),	510 lbs.	5.4 per cent. or	27.54 lbs. butter fat.
November,	1243 "	5.4 " " "	67.12 " " "
December,	1210 "	5.3 " " "	64.13 " " "
1894.				
January,	1160 "	5.3 " " "	61.48 " " "
February,	1090 "	5.4 " " "	58.86 " " "
March,	1029 "	5.2 " " "	53.51 " " "
April,	982 "	5.2 " " "	51.06 " " "
May,	869 "	5.3 " " "	46.06 " " "
June,	841½ "	5.4 " " "	45.44 " " "
July,	793 "	5.4 " " "	42.82 " " "
August,	711 "	5.6 " " "	39.81 " " "
September,	504 "	5.6 " " "	28.22 " " "
October (1-19),	276¼ "	6.1 " " "	16.86 " " "
	11218¾ "		602.91 " " "

FEEDING RECORD.

Daily Ration Oct. 20, 1893, to Jan. 10, 1894:
  Corn Ensilage, 30 lbs. Hay, 10 lbs. Bran, 8 lbs. Corn Meal, 4 lbs. Cotton-seed Meal, 1½ lbs. Oil Meal, 1½ lbs. Ground Oats, 2 lbs.

Daily Ration Jan. 11, 1894, to July 10, 1894:
  Corn Ensilage, 30 to 35 lbs. Hay, 2 to 3 lbs. Bran, 12 lbs. Corn Meal, 8 lbs. Cotton-seed Meal, 1 lb. Oil Meal, 1 lb.

Daily Rations July 11, 1894, to Sept. 1, 1894:
  Corn Ensilage, 30 to 35 lbs. Hay, 2 lbs. Bran, 10 lbs. Corn Meal, 6 lbs. Cotton-seed Meal, 1 lb. Oil Meal, 1 lb.

Daily Ration Sept. 2, 1894, to Oct. 20, 1894:
  Corn Ensilage, 30 to 35 lbs. Hay 1 to 1½ lbs. Bran, 8 lbs. Corn Meal, 3 lbs. Oil Meal, 1 lb.

Miss Bobolink 2157, A. G. C. C

Record for 1897, 12,436 lbs Milk, 696.46 lbs. Butter. Bred and owned at Millwood Farm, N. I. Bowditch, Proprietor, Framingham, Mass.

Millwood Farm Guernseys have been long and favorably known as being of large size, and showing great capacity and constitution. They are carefully handled and with good results. Nearly all were bred on the farm and descended from stock which Mr. Bowditch's father, the late E. F. Bowditch, imported and developed.

Miss Bobolink, as her picture shows, is typical of the character of Millwood Guernseys.

She has been a good and regular breeder. After calving, Sept. 21, 1896, she made the following record:

Gave from Oct. 1, 1896, to Oct. 1, 1897—12,437 lbs. milk.

Average test, 4.8 per cent. butter fat.

This equals 596.98 lbs. butter fat, or

696.40 pounds butter for 1 year.

In addition she has been milked three months longer, and gave from Oct. 1, '97, to Jan. 1, '98, 2480.5 pounds milk, which tested 5.3 per cent. butter fat or 153.37 lbs. butter, making the total record for the 15 months, 14,917.5 lbs. milk 849.83 lbs. butter.

The picture from which the half tone was made was taken Jan. 23, '98, or after 15 months' work, yet on that day she gave 21 lbs. milk. She was to calve in April, but just before time for calving, sickened with pneumonia and died.

Her milk record in detail is as follows:

		Milk Lbs.	Test Per cent.
1896.	October,	1,145	4.4
	November,	1,242.5	4.6 / 5
	December,	1,096	4.5
1897.	January,	1,036.25	
	February,	924.75	4.4
	March,	1,040	4.6
	April,	987.5	6.0
	May,	1,154.5	
	June,	1,049	
	July,	820.5	
	August,	991	4.6
	September,	950	
	Total for year	12,437	Average, 4.8
	October,	1,042	
	November,	748.5	
	December,	690	5.3
	Total,	14,917.5	

The tests with the Babcock were not made to correspond with the months' records, but were sufficient to obtain a good general average. The following statement is made as to her feed:

October 1st to May	Cob meal, 4½ qts.
	Bran, 4 qts.
	Oil meal, 1 pt.
	Hay, 16 lbs.
May to September,	The above grain and pasture.
September,	The above grain and 1 bush. cut corn fodder daily.
October to January,	The above grain and 16 lbs. hay.

She did not always have cob meal, when clear meal was fed, 1 quart less given.

Bred and owned at Ellerslie Stock Farm, L. P. Morton, Proprietor, Rhinecliff, N. Y.
Record 703 lbs. butter in one year.

Yearly Butter Records Over 400 Pounds.

Name & No. of Cow	Owner	Lbs. Milk	Per Ct. Butter Fat Range	Average	Lbs. Butter Fat	Lbs. Butter 1-6 Added to Fat	Age at Test
Imp Bretonne 3660	Levi P Morton	11,218.8	5.2 to 6.1		602.9	703.4	9 yrs
Buda 7178	"	8,559.0		6.96	602.7	703.0	6
Miss Bobolink 2157	N I Bowditch	12,437.0	4.4 to 6.0	4.80	596.9	696.5	13
Quibble 6017	Levi P Morton	10,548.0	4.3 to 5.8	5.17	545.3	636.0	6
Glenwood Girl 6th 9113	E T Gill	9,931.3	4.3 to 6.9	5.30		624.6	5
Imp Doutta Galla 4th 7675	Levi P Morton	9,166.0		5.66	519.0	605.5	6
Imp Villets Gem 8644	"	9,265.0		5.58	517.0	603.0	5
Fantine 2d 3730	Chas Solveson	9,748.0	5.0 to 5.6		516.6	602.7	8
Madge of Avon 6802	Levi P Morton	9,198.0		5.59	514.7	600.5	7
Yeksa's Queen 6631	A J Philips	9,886.6		5.20	514.1	599.8	5
Lily Ella 7240	Jas H Beirne	9,370.3	4.6 to 7.0	5.64	513.2	598.7	3
Imp Louise of Ellerslie 8930	Levi P Morton	8,770.0		5.81	510.0	595.0	2
Glenwood Girl 2d 9108	E T Gill	9,944.7	4.5 to 5.2			593.5	
Rutila's Daughter 6670	H McK Twombly	8,988.0		5.45	489.8	571.4	3
Imp Rydale's Jessie 7686	Levi P. Morton	8,980.0		5.43	487.5	569.0	6
Adario 5653	" "	7,378.0		6.61	487.6	569.0	7
Cousin Beatrice 7303	" "	10,023.0		4.83	484.1	565.0	5
Coralette 5722	Geo C Hill & Son	8,845.4		5.42	479.7	559.7	3½
Nubia's Vesta 5986 (for 6 mos.)	" " "	7,222.0		6.60	476.7	556.2	4
Imp May Jessie 7678	Levi P Morton	7,614.0		6.24	475.1	554.0	5
Imp Esmeralda 8657	" "	10,174.0		4.66	474.1	553.0	8
Quietness 6018	" "	8,299.0		5.62	466.0	543.7	6
Millwood's Perseverance	N I Bowditch	8,473.0		5.50	466.0	543.7	6
Glenwood Girl 3d 9109	E T Gill	9,273.7	4.5 to 6.2	5.07		541.8	7
Kings Myra 5339	Ezra Michener	8,611.0				539.5	4
Imp Beauty des Domaines 3d 4933	H W Comfort					535.4	6
Imp Honoria 4th 5989	H McK Twombly	9,589.0		4.75	455.5	531.4	5
Imp Rose of the Quartiers 8654	Levi P Morton	7,989.0		5.68	453.8	529.0	7
Imp Bonnie Cobi 7671	" "	7,474.0		5.91	441.7	515.0	5
Queen of Chenequa 6544	Chas Solveson	7,724.0		5.70	440.3	513.7	5
Imp Valentine De Grou 7641	Levi P Morton	8,267.0		5.32	438.8	512.0	7
Lady of Ellerslie 4543	" "	6,156.0	6.0 to 7.9	7.13	438.8	512.0	9
Nancy Marshall 5604	Ezra Michener					511.3	6
Teresita 8056	N K Fairbank				436.2	508.9	
Decasse 7669	Levi P Morton	7,208.0		6.04	435.4	508.0	4
Coca 4258	H McK Twombly	8,153.3		5.33	434.6	507.0	9
Morn 5947	N K Fairbank	7,606.8		5.71	434.3	506.7	3
Glenwood Girl 7th 9114	E T Gill	9,654.2	3.9 to 5.4	4.60		506.5	4
Glenwood Girl 4th 9110	E T Gill	9,906.7	3.8 to 5.3	4.43		504.5	6
Lauretta 3404	So Dak Agl Col	8,944.0		4.60	432.1	504.1	9
Imp Lady Archer 5th 6361	Levi P Morton	9,223.0	4.4 to 6.8	4.61	425.0	495.8	6
Madame Bishop 2724	Geo C Hill & Son	7,619.1		5.57	424.1	494.7	10

Yearly Butter Records Over 400 Pounds.—Continued.

Name & No. of Cow	Owner	Lbs. Milk	Per Ct. Butter Fat Range	Per Ct. Butter Fat Average	Lbs. Butter Fat	Lbs. Butter 1-6 Added to Fat	Age at Test
Sarepta Paulsdale 7387	So Dak Agl Col	8,477.0			424.0	494.7	3 yrs
Lulu of Haddon 10683	E T Gill	8,497.3	4.4 to 5.6	4.95		488.4	7
Imp Daisy 3d 7683	Levi P Morton	7,386.0		5.66	418.1	488.0	7
Champ's Sweet Briar 3d 3021	Minn Agl Exp Sta	7,678.3	4.8 to 7.5	5.43	416.8	486.3	11
Kindness 7234	N K Fairbank	9,447.4		4.41	416.6	486.3	4
Florence of Guernsey 6th 3215	Levi P Morton	8,809.0		4.82	414.6	483.7	12
Aldine 1211	Ezra Michener	8,615		4.92		482.8	13
Madame Tricksey 6510	Geo C Hill & Son	7,091.3		5.83	413.7	482.7	13
Deputy's East Lynne 6662	Levi P Morton	7,977.0		5.18	413.2	882.1	6
Queen Vashti 6651	A J Philips					482.0	7
Miss Caro 7304	Levi P Morton	7,101.0		5.80	411.8	480.5	5
Imp Supreme 3d 7639	H McK Twombly	7,054.0		5.17	411.2	479.7	2½
Jessie H 4348	Ezra Michener	8,653.0		4.86		479.3	7
Bonnie Beauty 5721	Geo C Hill & Son	7,168.4		5.77	410.5	478.9	5
Champion Snowflake 3086	A F Burke	9,055.0	3.2 to 5.2	4.50	407.5	473.5	7
Primrose Tricksey 7236	Geo C Hill & Son	7,317.8	5.0 to 6.8	5.50	406.9	474.7	3
Purity 2315	G H Davison	7,226.0				470.0	
Deanie 3d 7643	H McK Twombly	7,335.3		5.48	402.0	469.0	4
Marguerite of Springside 8544	Levi P Morton	6,732.0		5.94	399.8	466.5	5
Imp Trusty of Rhinecliff 8659	" "	8,679.0	4.4 to 5.3	4.60	399.0	465.5	6
Belvidera 7644	H McK Twombly	7,478.5		5.32	397.9	464.2	5
Young Dolly 3492	N I Bowditch	8,013.2		4.95	396.5	462.1	10
Avon Belle 7289	Levi P Morton	9,069.0		4.37	396.0	462.0	5
Fair Lad's Primrose 3244	Geo C Hill & Son	8,780.4		4.49	395.0	460.9	11
Pellet 6339	Levi P Morton	6,905.0	5.2 to 6.4	5.60	392.0	457.3	5
Coleta 7180	" "	6,486.0		5.93	391.3	456.5	5
Mindwell 6429	" "	7,380.0	4.5 to 6.0	5.24	391.0	456.1	6
Rosette Ford 3067	N Y Agl Exp Sta	6,501.1		6.00	390.1	455.1	5
Queen of Hearts 7121	H McK Twombly	7,395.0		5.26	387.0	453.5	2½
Imp Cesaree 2d 8649	Levi P Morton	6,647.0		5.85	385.9	453.0	5
Imp Royale 4th 8086	" "	7,277.0		5.35	385.0	452.7	5
Corona of Haddon 10684	E T Gill	7,680.1	4.2 to 6.1	5.41		451.2	5
Gladys of Doylestown 4243	Chas L Hill	7,200.0	5.0 to 6.5			450.0	9
Sarnietta 7248	Levi P Morton	7,110.0	4.5 to 6.3	5.40	384.0	445.0	5
Floret 6th 4313	Pa State College	8,416.0		4.57		447.0	7
Imp Rose of St Helene 2d 8082	Ed Severin Clark	6,336.0		6.00	381.0	444.5	9
Imp Minnet of Rhinecliff 8682	Levi P Morton	5,628.0		6.76	380.5	444.8	6
Imp Pomara 3d 1743	John C Higgins					443.0	14
Selectrina 6213	H McK Twombly	6,495.0		5.85	380.0	442.3	3
Maternalette 2d 6854	N K Fairbank	6,798.7		5.60	379.9	442.1	5
Countess Bishop 7869	Geo C Hill & Son	6,278.3		6.03	378.6	442.0	2
Prestoun 6570	" "	7,126.3	5.0 to 7.0			378.3	441.8	6
Select of Muster Hill 4061	Levi P Morton	7,651.0		4.95	378.0	441.0	7

Yearly Butter Records Over 400 Pounds.—Continued.

Name & No. of Cow	Owner	Lbs. Milk	Per Ct. Butter Fat Range	Per Ct. Butter Fat Average	Lbs. Butter Fat	Lbs. Butter 1-6 Added to Fat	Age at Test
Imp Egypta 8667	Levi P Morton	6,938.0	4.9 to 6.0	5.47	378.0	441.0	4 yrs
Duchinette 2d 7237	N K Fairbank	7,669.7		4.92	377.3	440.2	4
Acona 5652	Levi P Morton	7,362.0		5.25	376.5	439.4	7
Imp May Rose 4th 7682	" "	8,036.0	4.0 to 5.2	4.68	376.0	438.7	7
Imp Lily of the Spurs 4th 8391	" "	8,645.0		4.35	376.0	438.7	5
Grafin Von N 5708	N K Fairbank	6,550.0		5.74	376.0	438.7	3
Miss Chicago 6820	N I Bowditch	6,818.5		5.50	375.0	437.5	5
Melba 7618	N K Fairbank	8,319.3		4.50		436.8	
Nounon 6569	Geo C Hill & Son	6,590.0		5.68	374.3	436.7	6
Imp Rose of the Ville Amphrey 3d 7676	Levi P Morton	7,190.0	4.6 to 5.9	5.20	374.0	436.3	4
Comely 7642	H McK Twombly	7,279.0		5.13	373.5	435.7	6
Armorel 1st of Haddon 9529	E T Gill	7,775.0	4.1 to 6.5	5.30		435.1	
Robiline 7618	N K Fairbank	6,889.7		5.42	373.4	435.1	3
Careno 4999	N K Fairbank	6,200.1		6.00	372.0	434.0	4
Fernwood Robinette 3509		6,450.2		5.74	370.2	431.9	4
Bonnet 3488	N I Bowditch	6,618.8		5.60	370.1	431.8	10½
Coral of Paulsdale 3497	I J Clapp	7,294.8	4.7 to 5.75		369.2	430.7	6
Imp Tamar of Rhinecliff 8661	Levi P Morton	6,612.0		5.58	368.9	430.4	4
Mernie 4493	Geo C Hill & Son	7,160.0		5.13	367.8	429.1	4
Imp Guernsey Rose 5992	H McK Twombly	7,412.0		4.96	367.8	429.1	5½
Rachel of Ellerslie 7660	Levi P Morton	7,127.0		5.14	366.3	427.4	4
Suke of Rosendale 6520	Geo C Hill & Son	6,328.9		5.78	366.1	427.1	5
Regina's Trickey 5724	" "	6,520.0		5.61	366.0	427.0	5
Electra of Geneva 8059	N K Fairbank	7,146.6		5.11	365.0	425.8	2
Miss Cowslip 4662	N I Bowditch	7,927.0		4.60	364.7	425.5	8½
Oriole 4003	N Y Agl Exp Sta	6,659.7		5.46	363.8	424.4	4
Button 6010	Levi P Morton	6,810.0		5.28	363.6	423.7	6
Lady Bishop 6518	Geo C Hill & Son	6,586.2		5.51	363.1	423.6	2
Lucretia of Haddon 10831	E T Gill	6,502.8	4.9 to 7.8	6.31		423.5	4
Imp Esperance 5th of the Lohiers 7679	Levi P Morton	8,882.0		4.09	363.0	423.5	4
Imp La Pomare 8663	" "	7,393.0		4.90	362.0	422.3	4
Young Nacelle 7294	" "	6,367.0		5.71	361.3	421.5	4
Fernwood Roanoke 2d 8297	Geo C Hill & Son	6,573.5		5.49	360.9	420.1	6
Victorina 6212	H McK Twombly	6,551.0		5.48	359.0	419.0	3
Fernwood Roanoke 3903	Geo C Hill & Son	6,530.1		5.48	357.1	417.3	6
Primula 2d 8057	N K Fairbank	7,386.6		4.81	355.3	414.5	3
Arachine 8061	" "	8,726.1		4.06	354.3	413.3	3
Miss Bishop 7868	Chas L Hill	5,738.7			354.3	413.3	3
Rangoon 7247	Levi P Morton	7,440.0	5.5 to 7.3	4.76	354.0	413.0	3
Imp Flint 8392	" "	7,763.0	4.2 to 5.4	4.56	353.9	413.0	5
Hope Jewel 6801	" "	6,276.0		5.63	353.3	412.2	6

Yearly Butter Records Over 400 Pounds.—*Continued.*

Name & No. of Cow.	Owner	Lbs. Milk	Per Ct. Butter Fat Range	Per Ct. Butter Fat Average	Lbs. Butter Fat	Lbs. Butter 1-6 Added to Fat	Age at Test
Woglinde 5005	N K Fairbank	7,360.4		4.76	350.4	408.8	5
Lassie of Level Green 7967	H McK Twombly	5,825.5		6.00	349.5	407.8	3
Imp May Rose 5th 7681	Levi P Morton	5,953.0	4.0 to 7.4	5.84	348.0	406.0	3
Heiress 2d 5827	S C Hall	6,147.0		5.50		405.6	6
Miss Carrie 5290	Levi P Morton	6,908.0		5.03	347.4	405.4	8
Materna 1334	N K Fairbank	8,469.0		4.81	347.4	405.4	9
Young Celia 4513	N I Bowditch	7,387.5		4.70	347.2	405.0	8½
Alena 6099	Levi P Morton	7,317.0		4.72	344.9	402.4	7
Imp Rose of Lilyvale 3d 8645	Levi P Morton	5,679.0		6.05	343.6	400.9	4 yrs
Imp Gypsy of Natick 4435	" "	7,357.0	4.2 to 5.0	4.66	343.0	400.2	10
Imp Queen 4th 7646	" "	6,092.0		5.63	343.0	400.2	3½

Nubia's Vesta 5986, A. G. C. C.

Owned by Geo. C. Hill & Son, Rosendale, Wis. Record 7222 lbs Milk ; average test, 6.6 per cent. Butter Fat or 550 lbs Butter in 6 mos. Also 25 lbs 8 oz. Butter in seven days.

Bred and owned at Haddon Stock Farm, E. T. Gill, Proprietor, Haddonfield, N. J. A member of the noted Glenwood Girl family. Has made 593.5 lbs. butter in one year. Five daughters of Imp. Glenwood Girl have records of over 500 lbs. butter in one year.

Seven Day Butter Tests.

The following represents those butter records for a week exceeding 14 lbs., or 2 lbs. a day, that have been reported. There are undoubtedly many cows not in the list that should be. Such additions will be made when another edition of the Year Book is issued.

Name and Number.	Tester.	Lbs. Butter in Seven Days.			
Royalette, 3299,	F W Tratt,	28	lbs.	12	oz.
Nubia's Vesta, 5986,	Geo C Hill & Son,	25	"	8	"
Gully 5th, 1590,	S L Hoxie,	24	"	2	"
Lily Ella, 7240,	Jas H Beirne,	23	"	11¾	"
Bessie de la Palloterie, 1409,	E F Bowditch,	23	"	8	"
Fantine 2nd, 3730,	Chas Solveson,	23	"	7½	"
Fernwood, Lily, 1468,	L W Ledyard,	22	"	11½	"
Select, 2205,	Francis Shaw,	22	"	8	"
Kathleen, 38,	C M Beach,	22	"	4	"
Lilyeta, 7241,	Jas H Beirne,	22	"	1½	"
Forest Queen, 2019,	Corydon Peck,	22	"		
Duchess of Brittany, 1613,	S L Hoxie,	21	"	4	"
Rutila's Daughter, 6670,	H McK Twombly,	21	"	9	"
Blanche Charmau, 676,	A Warner,	21	"	14	"
Lucille, 115,	E F Bowditch,	21	"		
Garnet of Lehigh, 2208,	A F Fuller,	20			
Minnehaha, 1554,	Corydon Peck,	19	"	6	"
Elegante, 592,	L W Ledyard,	19	"	4	"
Polly of Fernwood, 1565,	L W Ledyard,	19	"	1½	"
Lady May, 531,	L W Ledyard,	19	"		
Fernleaf, 636,	L W Ledyard,	18	"	13	"
Bretonne, 3660,	Levi P Morton,	18	"	12½	"
Vestal of Larchmount, 1507,	Francis Shaw,	18	"	12½	"
Politesse, 1329,	Est E F Bowditch,	18	"	11	"
Mernalette, 5723,	N K Fairbank,	18	"	10	"
Stella 4th, 1598,	S L Hoxie,	18	"	4	"
Rosa of Maple Glen, 2459,	Corydon Peck,	18	"	2	"
Miss Bobolink, 2157,	Est E F Bowditch,	18	"		
Jessie of Lester Manor, 740,	Francis Shaw,	18	"		
Coraline, 1790,	I J Clapp,	18	"		
Coral, 2nd, 98,	Jas M Codman,	18	"		
Silky, 470,	Jas Logan Fisher,	18	"		
Belle Forest, 2nd, 68,	Jas P Swain, Jr,	18	"		
Valentine de Gron, 7641,	L P Morton,	18	"		
La Reine,	(Island cow),	18	"		
White Rose, 44,	F A & S W Comly,	17	"	11¾	"
Mary Marshall, 5604,	Ezra Michener,	17	"	12	"
Rosebud 4th of Les Vauxbelets, 1037,	I J Clapp,	17	"	10	"
Honoria 4th, 5989,	H McK Twombly,	17	"	10	"
Yeksa, 2426,	I J Clapp,	17	"	8	"
Primrose Tricksey, 7236	Geo C Hill & Son,	17	"	6½	"
Suke of Rosendale, 6526,	Geo C Hill & Son,	17	"	6	"
Virginia of Madison, 6000,	H McK Twombly,	17	"	4½	"
Miss Gypsy, 9641,	John W Scribner,	17	"	4	"
Gleenwood Girl 4th, 910,	E T Gill,	17	"	4	"
Lily of Prospect, 615,	I J Clapp,	17	"	4	"
Dawn, 711,	Francis Shaw,	17	"	2½	"
Coralette, 5722,	Geo C Hill & Son,	17	"	2	
Dolly's Ada, 9289,	G B Tallman,	17	"	1	
Treefoil, 522,	Chas R King,	17	"		
Merry Rose, 3766,	Ezra Michener,	17	"		
Berkshire Maid 2nd, 4600,	W D Richardson,	16	"	14½	"
Hazelnut, 1788	I J Clapp,	16	"	14	"
Madame Tricksey, 6519,	Chas L Hill,	16	"	12¾	"

Seven Day Butter Tests.—*Continued.*

Name and Number.	Tester.	Lbs. Butter in Seven Days		
Hermite,	(Island cow),	16	lbs.	12	oz.
Materna, 1534,	N K Fairbank,	16	"	8	"
Yeksa Queen, 6631,	A J Phillips	16	"	8	"
Countess Bishop, 7869,	Geo C Hill & Son,	16	"	7¾	"
Champion's Sweet Briar, 3rd, 3021,	Univ Minnesota,	16	"	7¼	"
Essence, 3667,	Levi P Morton,	16	"	4 6	"
Fanny, 410,	A Warner,	16	"	4	"
Musette Ford, 1600,	S L Hoxie,	16	"	4	"
Valencia, 664,	Jas Logan Fisher,	16	"	3⅓	"
Quartz, 4022,	Robt W Lord,	16	"	3⅓	"
Princessa 4th, 469,	Jas Logan Fisher,	16	"	2	"
Martha Lyons, 5697,	Wm H Caldwell,	16	"	1½	"
Coral of Paulsdale, 3497,	I J Clapp,	16	"		
Young Dolly, 3492,	E F Bowditch,	16	"		
France, 2207,	Francis Shaw.	16	"		
Flower, 640, F. S ,	(Island cow),	16	"		
Rose Forest 2nd, 888,	Corydon Peck,	16	"		
Daisy Pearl, 5990,	H McK Twombly,	15	"	15¾	"
Pancha, 205,	I J Clapp,	15	"	15½	"
Grand Daughter, 222,	E F Bowditch,	15	"	15½	"
Sweet Ada, 3596'	J M Eddy,	15	"	14½	"
Amanda, 222,	E F Bowditch,	15	"	12¾	"
Regina, 2691,	S C Kent,	15	"	12	"
Dolly, Ford 2nd, 1595,	S L Hoxie,	15	"	12	"
Nellie, 122,	Wm B Cooper,	15	"	11	"
Miss Bishop, 7868,	Chas L Hill,	15	"	10	"
Young Constance, 1415,	E F Bowditch,	15	"	8½	"
Primrose Ford, 1589,	S L Hoxie,	15	"	8	"
Glenwood Girl 6th, 9113,	E T Gill,	15	"	7¾	"
Coca, 4258,	H McK Twombly,	15	"	7½	"
Balboa Select, 7741,	Chas Solveson,	13	"	5¼	"
Rose des Catils 2nd, 3694,	Levi P Morton,	15	"	4	"
Darling of Braintree, 635,	Francis Shaw,	15	"	2	"
Selectrina, 6213,	H McK Twombly,	15	"	1½	"
Select 8th, 4059,	Francis Shaw,	15	"	1.1	"
Queenie Quartier, 1092,	I J Clapp,	15	"	1	"
Passagere 2nd, 1528,	Levi P Morton,	15	"	1	"
Worthy Beauty, 295,	T M Harvey & Son,	15	"		
Lily of Castel, 846,	I J Clapp,	15	"		
Victorine, 56,	Francis Shaw,	15	"		
Euphemia, 2005,	Ezra Michener,	15	"		
Regina's Tricksey, 5724,	Geo C Hill & Son,	15	"		
Cousin Bonita, 4512,	E F Bowditch,	15	"		
Miss Beautiful, 4598,	E F Bowditch,	15	"		
Glenwood Girl 3rd, 9109,	E T Gill,	15	"	15¾	"
Rosette Ford, 4067,	R L Hoxie,	14	"	14½	"
Lucretia's Daughter, 11256,	E T Gill,	14	"	14½	"
Richesse du Chene 5th, 7648,	H McK Twombly,	14	"	14	"
Queen Vashti, 6051,	A J Phillips,	14	"	12⅓	"
Moss Rose, 1180,	I J Clapp,	14	"	12	"
Margaret B, 2038,	I J Clapp,	14	"	12	"
Countess of Fernwood, 1464,	L W Ledyard,	14	"	12	"
Fleurie de Terte, 11336, G. H. B.	(Island Cow)	14	"	10	"
Belvidera, 7644,	H McK Twombly,	14	"	10	"
Brier Rose, 10316,	E T Gill,	14	"	9	"
Beauty of Geneva, 819,	N K Fairbank,	14	"	8	"
Beulah 2nd, 134,	J P Swain, Jr,	14	"	8	"
My Pet, 3094,	E N Howell,	14	"	8	"
Fantasia, S, 7921,	Chas Solveson,	14	"	8	"
Dainty's Maid, 5906,	G E Gordon,	14	"	8	"

Seven Day Butter Tests.—(*Continued.*)

Name and Number.	Tester.	Lbs. Butter in Seven Days.		
Fernwood Fancy, 37,	C M Beach,	14 lbs.	7	oz.
Guernsey Rose, 59992,	H McK Twombly,	14 "	6	"
Glenwood Girl 7th, 9114,	E T Gill,	14 "	5¾	"
Cloverdale, 9719,	John W Scribner,	14 "	5¼	"
Katie, 189,	J P Swain, Jr,	14 "	5	"
Glenwood Girl 2nd, 9108,	E T Gill,	14 "	4¾	"
Lenoxdale's Maid 5th, 5880,	Chas Solveson,	14 "	4½	"
Miss Jehan, 780,	Francis Shaw,	14 "	4	"
Primrose Ford 4th, 3302	S L Hoxie,	14 "	3½	"
Samanna, 11265,	E T Gill,	14 "	2¾	"
Sunbeam's Dawn, 9467,	Chas Solveson,	14 "	2½	"
Countess of Fernwood, 1404,	L W Ledyard,	14 "	2	"
Madame Bishop, 2721,	Chas L Hill,	14 "	2	"
Victorina, 6212,	H McK Twombly,	14 "	2	"
Yellow Gal, 11266,	E T Gill,	14 "	2	"
Vestal 2nd, 2210,	Francis Shaw,	14 "		
Select 2nd, 2229,	Francis Shaw,	14 "		
Bounty, 3091,	E N Howell,	14 "		
Sybil, 779,	Francis Shaw,	14 "		
Rose K, 199,	Chas R King,	14 "		
Betsey, 112,	E F Bowditch,	14 "		
Miss Agnes, 1026,	E F Bowditch,	14 "		
Virtue 6th,	(Island cow),	14 "		
Young Celia, 4513,	E P Bowditch,	14 "		
Good Morning, 3674,	Levi P Morton,	14 "		

Champion's Sweet Briar 3d 3021, A. G. C. C.
Property of Minnesota Agricultural Experiment Station.   Record 1803 lbs butter in one year.

Year's Herd Records. (Average per head.)

	Year.	Milk Lbs.	Butter Fat Tests	Butter Fat Lbs.	Butter Lbs.
Ezra Michener, Carversville, Pa.:					
15 cows of all ages until Jan. 14 when 1 heifer was added. Average milking period, 10 months	1897	6093.8			382.0
Entire herd (for over 10 yrs)					325.0
Levi P. Morton, Rhinecliff, N. Y.:					
62 cows and heifers	1892	6119.8			
83 cows and heifers	1896	5240.0	5.08		313.0
35 cows	1898	7089			514
Geo. C. Hill & Son, Rosendale, Wis.:					
12 head	1893	6199.6		327.6	382.2
Sydney Fisher, Knowlton, P. Q., Canada:					
19 cows—8 thoroughbred and 11 grades	1893				300.0
N. I. Bowditch, Framingham, Mass.:					
10 cows (average milking period, 9.7 months)	1896	6347.5	4.76	300.4	350.8
16 cows and heifers (average milking period, 10.75 months)	1897				397.1
H. McKay Twombly, Madison, N. J.:					
15 cows	1895	6626.0			406.4
14 cows	1897	6504.3	5.05		382.0
20 cows	1898	6490.0			396.4
28 cows	1898	6254.0			376.6

Owned by Jas. H. Beirne, Oakfield, Wis.

A Remarkable Record as a Two-year-old.

Lily Ella 7240 was born Oct. 19, 1893, and dropped a heifer calf March 12, 1896. She was bred April 9, 1897, to Springunde. Largest amount of milk given in one day, March 26, 1896, 29 lbs. 12 oz.; smallest amount of milk given in one day, March 17, 1897, 16 lbs. 3 oz. Highest single test, 7.90; lowest single test, 4.8 per cent. fat. The details are given in the following table:

Record of Lily Ella 7240, A. G. C. C.

Date	Milk	Test	Butter Fat	Butter	Daily Average Milk	Daily Average Butter	Days
March 18 to April 17	1178.31	4.9	57.74	67.36	38	2.328	31
April 17 to May 17	1007.44	5.0	50.37	58.76	33-9 oz	2.099	30
May 18 to June 17	914.00	4.95	45.24	52.78	29-8	1.824	31
June 18 to July 17	956.63	5.15	49.27	57.48	31-14	2.052	30
July 18 to Aug. 17	897.63	5.2	46.68	54.46	28-15	1.882	31
Aug. 18 to Sept 17	849.38	5.45	46.29	54.01	27-6	1.886	31
Sept. 18 to Oct. 17	695.38	5.45	37.90	44.22	23-3	1.575	30
Oct. 18 to Nov. 17	668.75	5.7	38.12	44.47	21-9	1.537	31
Nov. 18 to Dec. 17	611.44	5.9	36.07	42.09	20-6	1.50	30
Dec. 18 to Jan. 17	565.19	6.3	35.61	41.54	18-4	1.48	31
Jan. 18 to Feb. 17	546.88	6.63	36.37	42.43	17-10	1.466	31
Feb. 18 to March 17	479.00	7.00	33.53	39.12	17-2	1.597	28
	9370.3	5.64	513.19	598.72	25-6	1.761	

Island of Guernsey Butter Record.

Produce of a few dairy cows from January to December of each year:

    1894—Three Cows.   Butter, 1,029 lbs.

Doulta Galla III. 2549 P. S., dropped June, 1889 calved in October, 1894.
Richesse du Chéne II. 2958 P. S., dropped March, 1890, calved in April, 1894.
Glory of the Chéne 3437 P. S., dropped March, 1892, calved in August, 1894.

    1895—Four Cows.   Butter, 1,217 lbs.

Doulta Galla III. 2549 P. S., dropped June, 1889, calved in October, 1895.
Richesse du Chéne II. 2958, P. S., dropped March, 1890, calved in May, 1895.
Villa Rica 5740 P. S., dropped January, 1895, calved in February, 1895.
Richesse du Chéne V. 3499 P. S., dropped May 1893.

    1896—Four Cows.   Butter, 1,136 lbs.

Doulta Galla III. 2549 P. S., dropped June 1889, calved in October, 1896.
Richesse du Chéne II. 2958 P. S., dropped March, 1890, calved in March, 1896.
Villa Rica 3740 P S., dropped January, 1892, calved in March, 1896.
Glory du Chéne I. 4000 P. S., dropped September, 1894, calved in October, 1896.

    1897—Four Cows.   Butter, 1,217½ lbs.

*Doulta Galla III. 2549 P. S., dropped June, 1889, calved in October, 1896.
‡Richesse du Chéne II. 2958 P. S., dropped March, 1890, calved in April, 1897.
Richesse du Chéne VI. 3966 P. S., dropped May, 1895, calved in August, 1897.
Glory du Chéne I. 400 P. S., dropped September, 1894, calved in November, 1897.

    *Did not calve in 1897.   ‡Richesse II. sold about July.

In 1897 four cows the first six months, the other six months, three cows. This herd was the property of Mr. Jehan.

Madame Tricksey 6519 A. G. C. C.

First Prize and Sweepstakes, Guernsey Trans-Mississippi Exposition, Omaha, Neb., 1898. Owned by Geo. C. Hill & Son, Rosendale, Wis.

English Cows—Seven Day Records.

Nellie, 21.1 qts. milk; 2 lbs. 12 oz. butter.
Julia, 19.1 qts. milk; 2 lbs. 8½ oz. butter.
Lady, 17.1 qts. milk; 2 lbs 11 oz. butter.
Lady No. 2, 18.1 qts. milk; 3 lbs. 1½ oz. butter, at 11 years.
Lady No. 3, 16 qts. milk; 2 lbs. 8 oz. butter.
Lady No. 4, 14 qts. milk; 2 lbs. butter.
Lady No. 5, 17 qts. milk; 2 lbs. 9 oz. butter.
Lady No. 6, 16.2 qts. milk; 2 lbs. 9 oz. butter.
Lady No. 9, 17 qts. milk; 2 lbs. 15½ oz. butter.

Some Valuable Other Tests.

Lucille 115, 47 lbs. milk; 3 lbs. 1½ oz. butter in one day.
Lily of Alexandre 1059, 28 lbs. milk, test 7.2 per cent. of butter fat two months before calving
La Genista 2986, 110 lbs. milk; 6 lbs., 8 oz. butter in three days.
Bonnieline 2083, 2 lbs. 2 oz. butter in one day.
Nerrissa of Geneva 848, 2 lbs. 5 oz. butter in one day.
Maternalette 2127, 8 lbs. 8 oz. butter in three days.
Susanna 1817, 2 lbs. 2 oz. butter in one day.
Champion Beauty 1575, 2 lbs. 4 oz. of butter in one day.
Princessa 5th 2554, 98 lbs. milk; 6 lbs. 15 oz. butter in three days.
Valencia 664, 90 lbs milk; 7 lbs. butter in three days.
Margaret B 2038, 8 lbs. 7 oz. butter in four days, when two years old.
Tricksey 1760, 2 lbs. 7¾ oz. butter in one day.
Tricksey 2nd 2633, 8 lbs. butter in three and a half days.
Damsel, 39 lbs milk; 3 lbs. butter in one day,
Yeksa 2426, 112 lbs. milk; 10 lbs. butter in four days.
Poncha 2105, 4 lbs. 5 oz. butter in two days, when two years old.
Lily of Prospect, 615, 8 lbs. 10 oz. butter in three and a half days, when twelve years old.
Madeline 3rd 3685, 5 lbs. 4 oz. butter; 61½ lbs. milk in two days.
Amber 2nd 96, 2½ lbs. butter; 45 lbs. milk in one day.
Darling of Braintree 655, 7 lbs. 9 oz. butter in three and a half days.
Stella 4th 1598, rate of 18 lbs. 4 oz. butter in seven days.
Grand Daughter 222, rate of 15 lbs. 15½ oz. butter in seven days.
Dolly Ford 2nd, 1595, rate of 15 lbs. 12 oz. butter in seven days.
Young Constance 1415, rate of 15 lbs. 8½ oz. butter in seven days.
Primrose Ford, 1589, rate of 15 lbs. 8 oz. butter in seven days.
Betsey 112, rate of 14 lbs. butter in seven days.
Miss Agnes 1026, rate of 14 lbs. butter in seven days.
Champion Snowflake 3080, 10,357 lbs. milk in 13 months; 5.0 lbs. fat in three days.
Maternalette 2127, 8½ lbs. butter in three days.
Lily of Prospect 615, 8 lbs. 10 oz. butter in three and a half days.
La Genesta 2486, 6 lbs. 8 oz. butter in three days.

Fantine 2nd 3730, A. G. C C.
Record, 9748 lbs. Milk, 516.60 lbs. Butter Fat or 602 lbs. Butter.
Owned by Mr. Charles Solveson, Nashatah, Wis.

As her picture indicates, she is a well built dairy animal. Mr. Solveson writes: "The picture was taken when her udder was not more than two-thirds full and within a few weeks from calving." Fantine 2d was dropped

Sept. 12, 1887, making her at time of her record eight years old. She was sired by Imp. Pres. Garfield, No. 673, who was imported in dam, and was considered by his owner as one of the best Guernsey bulls. He dam was Fantine No. 2073, whose sire was the famous bull Imp. Lord Fernwood, No. 644, and Imp. Victory 2d of Larchmont No. 1511. Fantine 2d was bred by Alexander Scott of Ward, Pa., and has been owned since 1892 by Chas. Solveson of Nashotah, Wis. She is a cow of good digestive capacity, rugged constitution, and gives rich yellow milk.

From November 1, 1894, to October 31, Fantine 2d gave 9748 pounds of milk, which yielded 516.6 pounds of butter fat, which is equivalent, allowing one-sixth for the weigh of salt, water, etc., to 602.7 pounds of butter, or to 675.8 pounds of butter containing 80 per cent. fat. Her best day's yield was 47 pounds of milk, and her best month was 1318 lbs. milk, containing 69.85 lbs. of butter fat.

Her yield in detail is as follows:

Month	lbs.	testing		pounds fat	
November,	1260	lbs. testing	5.2	65.52	pounds fat
December,	1318	" "	5.3	69.85	" "
January,	1208	" "	5.2	62.81	" "
February,	1002	" "	5.6	56.11	" "
March,	1132	" "	5.3	59.99	" "
April,	990	" "	5.4	53.46	" "
May,	835	" "	5.3	44.25	" "
June,	555	" "	5.2	27.75	" "
July,	331	" "	5.	17.21	" "
August,	150	" "	5.6	8.40	" "
*September,					
October 6–31,	967	" "	5.3	51.25	" "
	9748			516,60	

*Was milked out a few times, of which no account was taken.

In this record there is one important thing to consider. Not only did Fantine 2d give this large amount of milk and butter, but it should be remembered that during the month of September and a few days in October no record was kept, as she was dry about four weeks, and on the 6th of October she gave birth to a fine bull calf. This makes the record one of eleven months instead of the full year.

During her year's work she was fed in the following manner: In winter, her daily ration consisted of 40 pounds of ensilage, 10 pounds cut stover, 8 pounds mixed hay, and for grain, 12 pounds of a mixture of oats, bran and dried brewers' grain, with two pounds of oil meal. The ensilage was rich in corn, and she was allowed in addition to this what wheat straw she cared to eat. On the 1th of April she was turned out to pasture with the rest of the herd, and her grain ration reduced to 6 pounds. The latter part of July she was fed soiling crops with the rest of the herd.

The Per Cent. of Fat and Total Solids in Guernsey Milk

In addition to the results reported in connection with the above records, there has been some other exceedingly interesting data of analytical work only.

From Mr. Levi P. Morton's herd a composite sample of mixed milk taken and analyzed by Prof W. W. Cook, then of the Vt. Ex. Station, showed 5.37 per cent. butter fat, 3.06 per cent. casein, and 15.18 per cent. total solids. Another instance from same herd was when 68 cows were tested, the results showing them all to be between 4 and 7 per cent.

In a competitive Home Test of Dairy Cattle, under the auspices of the Mass. Society for the Promotion of Agriculture, the Guernseys, belonging to Mr. Herbert Merriam, won first place, and the following shows the results:—

	Lbs. Milk.	Per cent Solids.	Lbs. Fat.	Lbs. Total Solids.
Polly of Concord 3909,	32.81	15.50	1.82	5.09
Golden Lily 5364,	37.75	15.03	1.97	5.68
Polly of Lincoln 2672,	37.06	15.32	2.00	5.69
Rose of Weston,	40.64	14.95	2.06	6.06
Weston Lily,	43.50	14.75	2.15	6.40
Average,		15.11		5.78
Totals,	191.76		10.01	

Mrs. Jos. Evans of Marlton, N. J., had samples of milk from six of his cows tested by Messrs. Marshall & Cockran of Philadelphia nearly every week for several months. The cow Arawana 2nd, ranged from 4.7 to 6.9 per cent.; Juno Princess, from 4.4 to 6.1 per cent.; Rouvots, from 3.9 to 5.6 per cent.; Comonnie 2d, from 4.0 to 6.7 per cent.; Maid of Canterbury, from 3.6 to 5.3 per cent.; and Imp. Lady Margot, from 4.0 to 5.5 per cent.

August 28, 1897, the herd of Guernseys owned by Mr. W. D. Richardson of Garden City, Minn., was officially tested by the State Dairy Inspector, who found the following results :—

	Lbs. Milk.	Per cent. Butter Fat.	Time since Calving	Age
Myrtle Maid 9065,	16.0	7.7	12mos.	2yrs.
Hazelwood Select 9062,	18.0	5.4	11	3
Jennie Select, 6009,	30.0	6.4	5	6
Imp. Ophir 2520,	25.0	5.6	5	12
Hartha Fanchion 11209,	18.0	5.8	5	3
Berkshire Maid Select 6340,	24.0	6.4	6	7
Martha of Hazelwood 11210,	20.0	6.2	6	2
Little Milk Maid 9066,	12.0	7.0	8	2
Berkshire Maid 2130,	30.0	5.3	5	12
Jennie Lewlissa 9063,	25.0	6.4	6	3
Milk Maid 2nd 5225,	22.0	5.4	8	6
Average,	22.0	6.1		4½

Test includes morning and evening milk.

Butter from this herd has received the highest awards at the State Dairy exhibits, and at the World's Fair in 1893. It has been highly spoken of on account of its fine natural color and flavor.

Mr. Chas. H. R. Triebels writes of his herd of Grade Guernseys:—

"I have had the milk from nine of my cows tested by Messrs. Marshall & Cockran, No. 215 North Fifth Street., Philadelphia, Pa., and give the result below— and would state that the cows were milked and the milk well stirred previous to the samples being taken. First samples Feb. 28, 1898:—

	Specific gravity.	Per cent. of fat.	Per cent. of solids not fat.	Per cent. of total solids.
Cow No. 1,	35.0	5.60	9.80	15.40
" 2,	33.4	5.70	9.40	15.10
" 3,	36.5	6.00	10.24	16.24
" 4,	37.2	7.50	10.69	18.19
" 5,	34.8	6.00	9.80	15.80
" 6,	33.7	3.80	9.48	25.28

Second Sample March 1, 1893.

Cow No. 7,	34.9	6.10	9.86	15.96
" 8,	34.5	6.90	9.90	16.80

Third Samples March 3, 1898.

Cow No. 9,	34.8	7.00	10.03	17.03
" 1,	34.3	5.60	9.63	15.23
" 3,	36.3	6.20	10.20	16.40
" 4,	34.8	7.60	10.10	17.70
" 5,	32.2	5.40	9.04	14.44
" 6,	32.8	3.10	9.17	14.27

I have had registered Guernsey bulls for nearly twenty years, and always getting the best to be had, so that in my stock there is the blood of six different strains. The last bull I purchased is from the Houston estate, Chestnut Hill, Philadelphia, in May last; and you will notice that he was registered. My cow No. 4 has had two calves, and I think she is hard to beat. Perhaps you may think that the cows were fed extra to produce the results stated; but I can assure you that they were fed as usual.

Mr. Triebels' herd is at "Rubicam Farm," Willow Grove P.O., Montgomery Co., Pa.

Rutila's Daughter 6670, A. G. C. C.

The property of Mr. H. McK. Twombly, Madison, N. J.

Rutila's Daughter was bred by Mr. Francis Shaw, Wayland, Mass., and dropped Oct. 9th, 1891. Her sire was Picotte's Squire, No. 1991 (by Imp. Squire of Larchmont 911, out of Imp. Picotte 2d 2218). Her dam was Rutila, No. 2912, by Tobasco 451; grand-dam Imp. Rita 2d 784.

During season of 1895 Rutila's Daughter won every prize for which she was entered, viz. :—

New York State Fair, Syracuse,	{	First prize in class 3 years or over. Sweepstakes as best female over 2 years. In first prize herd.
Bay State Fair, Worcester,	{	First prize in class between 3 and 4 years old. In first prize herd.
Inter-State Fair, Trenton, N. J.	{	First prize in class between 3 and 4 years old. In first prize herd.
Live Stock Society' Show, New York,	{	First prize in class 3 years or over. In first prize exhibition herd. In special silver cup herd. Championship as best female.

In 1897, 1st prize cow, New York State Fair. Sweepstakes cow, New York State Fair. In 1st prize herd, New York State Fair. 1st prize cow, New Jersey State Fair. In 1st prize herd at New Jersey State Fair.

BUTTER RECORDS.

1894 from Mar 15 to Dec. 31 (first calf),	7553⅔ Lbs. Milk.	5.26% Fat.	466.18 Lbs. Butter.	Containing 85% Fat
1895,	9988 "	5.45 "	576.26 "	"	"	"
1896,	9042¾ "	5.39 "	573.39 "	"	"	"
1897,	7554½ "	5.85 "	466.58 "	"	"	"
1898,	7545½ "	5.36 "	384.70 "	"	"	"

It should be borne in mind the cow went the round of the Fairs in 1895 and 1897 and had milk fever in 1896. Her daughter, Sheet Anchor's Rutila 9170, dropped first calf July 16, 1898, and gave from July 20 to Dec. 31, 1898, 3905¾ lbs. milk, 4.80 per cent. fat, 220.54 lbs. butter, containing 85 per cent. fat. She gave in January, 1899, 44.14 lbs.; February, 38.62 lbs.

Cream Produced by the Breeds.

One of the interesting calculations made at the Geneva, N. Y., station test of the breed is the amount of cream produced by them. These figures will be of especial interest to those who sell cream. We therefore give a summary of the data supplied. It seems that the Ayrshires produced 580.1 quarts at the average of the four years period, the Devons 434.3 quarts, the Guernseys 676.5 quarts, the Holsteins 630.5 quarts, the Jerseys 668.5 quarts, and the Shorthorns 637.4 quarts.

The amount of milk required to produce a pound of cream was in the order of the names above given, 5.58 pound., 4.35 pounds, 3.80 pounds, 5.95 pounds, 3.60 pounds and 4.50 pounds.

The cost of food in cents for a quart of cream being for the Ayrshires 8.5, Devons 8.63, Guernseys 6.82, Holsteins 8.04, Jerseys 6.79 and the Shorthorns 7.26.

The value of the cream per cow being in the order of the names above given, $116.02, $86.86, $135.27, $126.10, $133.70 and $127.48.

The director figures out the profits for cream selling as follows: For the Ayrshires $65.48, Devons $48.44, Guernseys $87.70, Holsteins $74.04, Jerseys $86.80, Shorthorns $79.92.

Guernseys as Economical Butter Producers.

IMPORTANT TRIALS AT THE AGRICULTURAL EXPERIMENT STATION FAVORABLE TO THE GUERNSEY.

Perhaps no breed has as honestly won its high rank as butter producers as have the Guernseys. Never forced for high records, they have stood upon the work they would do at the pail or churn. It is especially gratifying to notice how they are received in the sections where they have been introduced. Go into New England, down the Hudson in New York, into Eastern Pennsylvania, Delaware, New Jersey and Wisconsin, and you find not only fine herds of thoroughbreds, but you will notice that the dairymen of those sections have been impressed with their substantial, business-like appearance and golden colored products, and have been drawing on the breed for the grading up and improving the dairy stock of their section.

The Guernsey as a dairy cow has been more talked about since the World's Fair than she ever was before. It is undisputed that the Guernsey butter has the richest natural color of that of any breed. The Guernsey the world over has the rich, yellow skin, which old-time dairy people always said indicated a good butter cow. The prevailing Guernsey colors are white and bright red, shading into fawn color.

The most remarkable characteristic of the Guernsey is the *richness* of the animal. This richness is combined with good size and constitution, and particularly quiet, gentle and tractable temperament. They are rangy, deep animals, characterized by plenty of udder development, with soft, silky texture of skin and creamy color. As a four-year-old they are usually giving 16 to 18 quarts of milk during the first four or five months after calving, with ordinary keep, and are persistent milkers. Their milk and cream is of marvelous color and richness, and butter that in grain, flavor and golden color excels that of any other breeds.

Their ability to produce butter fat and butter at a low cost demands the most careful attention of dairymen. At the New York experiment station, several of the dairy breeds are being carefully tested. The annual report of the directors, which was recently issued, gives the result of the first two periods of lactation. In both instances, the Guernseys produced butter fat at the least cost, as the following shows:

COST OF BUTTER FAT PER POUND.

	1st Period.	2d Period.
Guernsey,	18.4 cts.	15.6 cts.
Jersey,	20.0 "	18.5 "
Devon,	23.0 "	19.0 "
Ayrshire,	24.3 "	24.8 "
Am. Holderness,	26.3 "	22.8 "
Holstein Friesian,	26.3 "	26.4 "

This agrees with the work done at the New Jersey experiment station, and with the average results of the World's Fair.

Cost Per Pound of Butter Produced.

	New Jersey.	World's Fair.
Guernsey,	15.3 cts.	13.1 cts.
Jersey,	17.9 "	13.3 "
Ayrshire,	20.6 "	
Shorthorn,	20.8 "	15.8 "
Holstein,	22.4 "	

New Jersey Experiment Station Report for 1889-90.

Mature Cows. Tested Nearly Nine Months.

	Avge. lbs. milk daily.	Daily Avge. butter fat.	In 300 days.		Profit.
Three Guernseys,	19.10	0.97 lbs.	291 lbs. at 25c.,	$72.25	
Cost of food per lb. of fat, in cents,			.1530	44.52	$28.23
Three Jerseys,	18.45	0.90 lbs.	270 lbs. at 25c.,	67.00	
Cost of food per lb. of fat, in cents,			.1790	48.33	18.67
Three Holsteins,	24.27	0.86 lbs.	258 lbs. at 25c.,	64.50	
Cost of food per lb. of fat, in cents,			.2240	57.79	6.71
Three Ayrshires,	19.76	0.73 lbs.	219 lbs at 25c.,	54.75	
Cost of food per lb. of fat, in cents,			.2260	45.11	9.64
Three Short Horns,	19.56	0.74 lbs.	222 lbs. at 25c.,	55.50	
Cost of food per lb. of fat, in cents,			.2080	46.18	9.32

New York Experiment Station, 1891-92.

Young Heifers. Their First Season. Avge. Ten Months.

	Avge. lbs. milk daily.	Daily avge. butter fat.	Avege. lbs. of butter (not fat), 10 mo.		Profit.
Two Guernseys,	16.00	0.90	268 lbs. at 25c.,	$67.00	
Cost of food per lb. of butter, in cents,			.1470	39.40	$27.60
Three Jerseys,	14.07	0.89	267 lbs. at 25c.,	66.75	
Cost of food per lb. of butter, in cents,			.1670	44.58	22.17
One Holstein,	22.65	0.79	241 lbs. at 25c.,	60.25	
Cost of food per lb. of butter, in cents,			.2204	54.50	5.75
Four Ayrshires,	18.40	0.61	188 lbs. at 25c.,	47.00	
Cost of food per lb. of butter, in cents,			.2303	43.33	3.67
Two Holderness,	13.40	0.52	157 lbs. at 25c.,	39.25	
Cost of food per lb. of butter, in cents,			.2204	34.60	
Two Devons,	12.65	0.51	152 lbs. at 25c.,	38.00	
Cost of food per lb. of butter, in cents,			.2287	33.70	4.30

This shows the Guernseys to be the most economical producers of butter; and such golden yellow butter, too! There is no mottled color to it. This true, golden, cow color is the most attractive feature on the market. It is truly said that the Guernseys have but to be tried to be appreciated.

Purity No. 2315, A. G. C. C.
Owned by Dr. G. Howard Divison, Millbrook, N. Y., A noted prize winner.
Ranking Guernsey in Thirty Days Butter Tests at Worlds Fair in 1893.

Guernseys at the World's Fair.

It is hoped that the results of the Columbian Dairy Test at the World's Fair, in 1893, will not be misjudged by the friends of the Guernseys. However pleasing it might have been to the breeders to see their favorites rank first in all particulars, the conditions of the tests and the difficulty encountered in securing animals for the trial did not warrant it.

With a total of about 6,000 cows, living and dead, in the Register the Guernsey breeders found themselves seriously hampered is making selections for cows to send to Chicago. The American Guernsey Cattle Club, being a younger organization than either the Jersey or Shorthorn Clubs, they were unable, through lack of sufficient funds, to make as thorough a search for cows as these other organizations, who had salaried agents travelling and selecting animals for the work for several months previous to the opening of the tests. The Guernsey Club was forced to send to Chicago such cows as public spirited breeders were willing to loan for the work. Many of the best Guernseys were not fresh at the proper time, and it is regretted that several were not available through the unwillingness of owners to risk their valuable animals under such conditions as the tests required. When the time came for announcing the twenty-five, the Guernsey Club had but twenty-six available to select from, while the Jerseys had fifty-three. Despite the disadvantageous circumstances, the Guernsey Club deemed it best to be represented, and used all possible means to make a creditable showing.

Particular attention is called to the peculiar rules which were arranged at the meetings where other breeders had influence strong enough to secure them to their advantage. The rules called for the rating of the butter at prices to be determined by the scoring of same, by a jury of three, and fixed the price much higher than that realized by the American farmer. The scale of points upon which the butter was judged was to the detriment of the Guernseys. It consisted of 55 points on flavor, 25 on grain, 10 on solidity, and 10 on color. This gave a total of 35 points on grain and solidity, shutting out salt, which was claimed to be governed by the skill of the butter-maker. That this is true to a large extent we will not question. At the same time, judging from the high score of the Guernsey butter on flavor, and that the butter being less solid dissolved the salt more evenly, it is believed it would have received higher commendation had it being submitted to the jury on that point.

The thirty-five points on grain and solidity are more than is common in scale of points usually used in judging butter. This gave the Jerseys an advantage. On this point the faulty construction of the dairy house was a great drawback to the softer butters. The glass roof admitted heat to such an extent that it was impossible to hold the temperature at a point conducive to high quality of butter. On the other hand, the higher natural color of the Guernsey butter was marked, but in none of the rules was natural color given any credit over that artifically colored. The Guernseys could not receive a just scoring on same, as the rules allowed coloring matter to be used at the discretion of the superintendents of the breeds competing. This allowed the other breeds to artificially imitate the natural color of Guernsey butter, depriving this test of any value, while admitting the superiority of Guernseys in this particular.

All these helped give the Jerseys a higher score, and consequently greater price for their butter, when in the open market it would not have sold any higher; the price being so much greater than that received by the farmers in open market, that a small fraction of a pound would make a preceptible money difference and pay for considerable feed.

It should be noted by all friends of the Guernseys that on flavor the Guernsey butter was ahead, and would have had much advantage on color, but for the unfortunate rules.

The rules also called for charging or crediting the cows with their gain or loss of live weight. This is something the intelligent farmer knows is never taken into consideration in their business. It should have never been made one of the conditions of the contest. It allowed the feeders to fatten their animals, and receive pay for the extra feed the animals received, yet did not convert into dairy products.

Sweet Ada No. 3956.

Owned by J. M. Eddy, Saratoga Springs, N. Y. Ranking Guernsey in Cheese Test, and Sweepstakes Guernsey in all Tests, Columbian Dairy Tests, Chicago, 1893.

The Results.

The results of the tests bear several interpretations. All that has been written comes from partisan sources. There is little doubt that the Jerseys as a breed rank first according to the rules for the conduct of the test. The Guernseys follow. If the rules are thrown aside, and the data discussed under the conditions the dairymen of the country are placed and come in everyday contact with, the results are changed. In the cheese work, the Guernseys have the two leading cows. In the butter test, they are represented by two and three of the best five cows of any breed.

In cost per pound of products the Guernseys won, as is seen by the following:

1st Test Cheese,	6.74 cts.	6.76 cts.	11.31 cts.
2d Test Butter,	13.4 "	13.3 "	15.9 "
3d Test Butter,	12.8 "	13.3 "	15.8 "
Average for Butter,	13.1 "	13.3 "	15.85 "

The Chicago test has dismissed into history the enormous tests made so public in the past. The Guernsey has met her cousins in such a manner that the work points out to the dairyman that the dairy cow is an individual, and that the Guernseys and the Jerseys both contain such animals, and rightly claim superiority as dairy breeds.

Twice has the Guernsey produced butter at the least cost in trials at the experiment stations under equal conditions. Here at Chicago she divided the first place with her neighbor, and beat her on cost of butter and cheese per pound.

Nor in the least is it to be regretted that the breed entered the contest. By so doing, they have made friends and received public attention that will be beneficial in the future.

Agreeing with this work at Chicago, we find in the reports of the Directors of the New York Station the results of trials during two periods of lactation of several of the dairy breeds, the results of which is given on another page.

These records seem to show conclusively that the Guernsey and the Jersey are the leading breeds for butter production, and we feel will bear out our claim that the Guernseys produced more cream and made more profits in selling cream than any of the other breeds. It is truly said that the Guernsey has but to be tried to be appreciated. Another advantage of the Guernsey cow is in the size of her calves. The farmer who desires to improve his dairy animals by crossing will select the Guernsey bull. The calves will have greater size and be more profitably turned into veal, or if raised, will take from the potency of the Guernsey the marvelous richness of the breed. All that is needed to popularize the Guernsey cow among practical farmers in any section is the possession of a single herd. A few visits and careful watching of such a herd will convince all dairymen of the value of Guernseys.

In connection with this article we give half tone engravings of the leading Guernsey in each of the three tests.

Materna No. 1334.

Owned by N. K. Fairbank, Chicago, Ill. First premium in show ring for Guernseys at the World's Fair. Also best Guernsey in the Ninety Day Butter Test at the World's Fair

The likeness of the grand cow "Materna" 1334 hardly does the cow justice. Her large business looking udder with tortuous milk veins are not shown to advantage. A good idea can be formed however of the great depth of body indicating the power and capacity for whioh Materna is noted.

Materna was sired by Imp. Amber No. 145; dam Imp. Duchess of Geneva No. 847. She was dropped on Sept. 3d, 1882, and is thus 16 years old, yet she carries with her that freshness and health not found in many a younger animal.

Materna 1334 is owned by Mr. N. K. Fairbank, and kept at his beautiful summer home at Lake Geneva. She has attracted the attention not only of all *Guernsey* breeders, but every admirer of a dairy cow, by the great work she did at the World's Fair Dairy Test, and in the Show ring. Not only was she accorded first place among the aged cows, and as sweepstake cow, but she was one of the leading ones in all three of the Tests at Chicago. In the first test, the cheese test, she ranked third among the Guernsey herd and 18th among the 75 cows. In the second of the ninety-day butter test, she was the leading Guernsey and 9th among the entire 73 cows. In the thirty-day butter test she stood 5th among the Guernseys and 15th among the entire number.

Her work is shewn by the following:

Cheese test,	597.2 lbs. milk.	62.01 lbs. cheese.	Net profit $ 4.82
90 Day butter test,	3511.8 " "	153.39 " "	" " 57.82
30 " " "	1058.4 " "	54.68 " "	" " 17.54

In one hundred and fifty one days she gave 5167.4 lbs. of milk an average of over 1030 lbs. of milk a month, or over 15 quarts a day.

From this milk was made 62 lbs. of cheese and 208 lbs. of butter, or averaging 3 lbs. of cheese while cheese was made, and 1.73 lbs. of butter while butter was made, for 150 days coutinuous work amid all the difficulties to be encountered at such a place as the Fair grounds were.

Aside from the amount of cheese and butter given, it is interesting to note that after deducting the cost of food there was shown a net profit for the period of $80.18 or 52½ cents a day.

This remarkable record was made under rules largely favoring the interests of another breed. Had it been under such conditions as the average dairymen are placed she would have made a better showing.

PRIVATE HERD BOOKS.

New editions. Reduced in price. Conveniently arranged to preserve identity of animals. Specially adapted for Guernsey Breeders.

Each volume is tastefully made up, and gives a chance to preserve the pedigree, sketch, and breeding record of each animal, and also an opportunity to keep the milk and butter fat record.

The importance of maintaining a home record of a breeder's herd can not be over estimated. Advancing dairy knowledge demands of the breeder some definite statement of the breeding and dairing qualities of his animals. Then by retaining a copy of the sketch of the animals as sent to the office of the club at time of registration the owner can always identify his cattle. This is especially desirable when one does not come in daily contact with his animals, or in settling estates.

PRICES POSTPAID.

For 50 animals, bound in half Russia	$2.50
For 100 animals, bound in half Russia	3.50
For 100 animals, bound in half Russia	5.50

POCKET HERD MEMORANDUM BOOK.

Contains place for pedigree, breeding and milk and butter records; also, bull service.

Book complete, seal leather	$1.25 postpaid
Extra pads	25 cents "

The covers will outwear many pads, and new ones can be inserted.

CALF REGISTERS.

Specially arranged to preserve the sketches and pedigrees of 100 animals, thus giving breeders a ready means for identifying their young stock at all times, and preserving their larger books for breeding animals. Price only $1.50 postpaid.

SAMPLE PAGES SENT ON APPLICATION.

SCALE OF POINTS FOR GUERNSEY COWS.

ADOPTED BY THE AMERICAN GUERNSEY CATTLE CLUB.

	Points.		Counts.
Quality of Milk.	30	Skin deep yellow, in ear, on end of bone of tail, at base of horn, on udder, teats, and body generally,	20
		Skin loose, mellow, with fine, soft hair,	10
Quantity and Duration of Flow.	40	Escutcheon wide on thighs, high and broad, with thigh ovals,	10
		Milk veins long and prominent,	6
		Udder full in front,	6
		Udder full and well up behind,	8
		Udder large, but not fleshy,	4
		Udder teats squarely placed,	4
		Udder teats of good size,	2
Size and Substance.	16	Size for the breed,	5
		Not too light bone,	1
		Barrel round, and deep at flank,	4
		Hips and loins wide,	2
		Rump long and broad,	2
		Thighs and withers thin,	2
Symmetry.	14	Back level to setting on of tail,	3
		Throat clean, with small dewlap,	1
		Legs not too long, with hocks well apart in walking,	2
		Tail long and thin,	1
		Horns curved and not coarse,	2
		Head rather long and fine, with quiet and gentle expression,	3
		General appearance,	2
	100		100

For Bulls, deduct 20 counts for udder.

For Heifers, deduct 20 counts for udder.

King's Myra, No. 5339.
Property of Ezra Michener, Cottage Farm, Carversville, Pa.

King's Myra, No. 5339, another four years old cow of Mr. Michener's, won the first prize of $100 given by the Guernsey Breeders' Association one year ago. She gave in the year 8611 pounds of milk that yielded 539½ pounds of butter.

Constitution of The American Guernsey Cattle Club.

Adopted at the Annual Meeting, Dec. 19, 1883.
Amended Dec., 1896.

PREAMBLE.

We, the undersigned, breeders of Guernsey Cattle, recognizing the importance of a trustworthy Herd Book, that shall be accepted as a final authority on all questions of Pedigree, and desiring to secure the influence and co-operation of those who feel a genuine interest in jealously guarding the purity of this stock, do hereby agree to unite in forming an Association for the publication of a Herd Book, and adopt for our government the following Constitution:

ARTICLE I.

This Association shall be styled The American Guernsey Cattle Club.

ARTICLE II.

The members of this Club, for Life, shall comprise the present *members* and such other persons as may be admitted, as hereinafter provided.

ARTICLE III.

The officers of the Association shall consist of a President, two Vice-Presidents, a Treasurer, and a Secretary. The President, the Treasurer, and the Secretary, together with *seven* members of the Club, *all chosen by ballot*, shall constitute an Executive Committee, with power to manage the affairs of the Club, subject to the provisions of the Constitution; fix the location of its headquarters; prescribe the manner in which its business shall be transacted, *and appoint other members of the Club* to fill vacancies which may occur in the Committee *during the year. At the ensuing annual meeting such vacancies shall be filled for the remainder of their unexpired term by vote of the Club.*

ARTICLE IV.

The annual meeting of the Club shall be held on the *second* Wednesday in December of each year, at such place as shall be designated by the Executive Committee (of which notice shall be sent to the members at least one month previously), for the election of officers and members of the Executive Committee, *the transaction of business*, and the discussion of questions of general interest to the Club.

The President, Vice-Presidents, and the Secretary and Treasurer shall be elected annually. The seven members elected as the executive committee shall draw by lot terms of one, two, three, and four years, two members for one year, two members for two years, two members for three years, one member for four years, and as their terms of office expire, members shall be elected annually in the same order and number to fill their places, for the term of four years each.

At all meetings of the Club members may vote in person *or by proxy. Proxies must be given to members only, shall only be good at the meeting for which they were given, and must be duly authenticated on forms adopted by the Club.*

ARTICLE V.

Each applicant for membership shall be recommended in writing by one or more members of the Club, as a reliable and careful breeder, *on a form furnished by the Secretary; such application shall first be presented to a Special Committee of three members, appointed annually by the President, upon whose recommendation* the Secretary shall send the names of the applicant and of the member or members recommending him, to all the members of the Club, and request their ballots for or against such applicant within thirty days of said notice. Three negative votes shall reject. The Secretary shall notify candidates of their election, and hey will be admitted as members on signing the Constitution and paying the entrance fee. The failure to do this within sixty days after the date of the above notice by the Secretary may be regarded as a forfeiture of election.

ARTICLE VI.

Each member shall pay an entrance fee of Fifty Dollars. *All moneys received shall constitute a fund* to defray the expenses of publishing the Herd Book, and other charges incidental to the transaction of the business of the Club, *under the direction of the Executive Committee.* No officer or member shall be authorized to contract any debt in the name of the Club; all its transactions shall be for cash.

ARTICLE VII.

The Herd Book shall be edited by the Secretary, under the immediate control and supervision of the Executive Committee, and shall be published only with its official approval. The price at which it shall be sold to members and others shall be determined by the Committee.

The Secretary shall keep on file all documents constituting his authority for pedigrees, and shall hold them subject to the inspection of any member of the Club.

ARTICLE VIII.

Should it occur, at any time, that any member of the Club shall be charged with willful misrepresentation in regard to any animal bred or owned by him, or with any other act derogatory to the standing of the Club, the Executive Committee shall examine the matter, and if it shall find that there is a foundation for such a charge, the offending member may be expelled by a vote of two thirds of all the members of the Club.

ARTICLE IX.

This Constitution may be altered or amended at any annual meeting of the Club, by the assent of three fourths of the votes cast, sixty days previous notice of the amendment having been sent by the Secretary to all of its members.

Imp. Rose of St. Helene 2d, A. G. C. C.
Owned by Ed. Severin Clark, Cooperstown, N. Y. Record, 447 lbs. Butter in one year.

By-Laws of The American Guernsey Cattle Club.

1. Special meetings of the Club may be called by the President, or by the Executive Committee, at the written request of ten members. Twenty days' notice is required. The object of the meeting shall be announced in the call, and no other business shall be transacted at that meeting.

At the annual meeting one eighth, at a special meeting one fourth, of the members of the Club shall form a quorum.

Five members of the Executive Committee shall form a quorum.

2. An Auditing Committee of one or more members of the Club shall be nominated by the President at each annual meeting to audit the accounts of the Treasurer before they are submitted to the next annual meeting.

3. All entry fees are payable in advance. Firms or partnerships cannot benefit by member's fees unless each partner is a member.

The charges for registry of Guernsey cattle shall be as follows:

For Animals Owned and Entered by Members of the Club.

Home-bred animals under six months of age,	$1.00
Imported animals of any age, if offered within six months after landing,	1.00
Home-bred animals over six months of age,	2.00
Imported animals of any age, offered after six months after landing,	2.00

For Animals Registered by Non-Members.

Home-bred animals under six months of age,	$2.00
Imported animals of any age, if offered within six months after landing,	2.00
Home-bred animals over six months of age,	3.00
Imported animals of any age, if offered after six months after landing,	3.00
Transfer Fee, in all cases, for each animal,	$1.00

5. APPLICATIONS FOR IMPORTED ANIMALS.—No animal hereafter imported shall be entered in the Herd Register of the American Guernsey Cattle Club, unless previously registered in a Herd Register on the Island of Guernsey or in the Herd Register of the English Guernsey Cattle Society.

6. With imported Guernseys must be certificates of breeding and sale on the forms furnished by the Secretary of this Club. On the back of each certificate must be a sketch of all white markings of each animal, made before shipment, and certified to by the breeder or seller. There shall also be stated therein the name and residence of the breeder, the name and residence of the importer, the date of birth of the animal, the sex, the Island or English Herd Book name and registered number, and the Island or English Herd Book name and registered number of the sire

and of dam of the animal offered for entry, if the same are registered, as they must be if living and on the Island at the date of execution of such certificate. If the imported animal be in calf, the owner of the serving bull shall affix his name to the breeding certificate with name and registered number of the serving bull and date of service.

These signatures and statements of the breeder or seller shall be forwarded to the Secretary of the Island or English Herd Books in which the animals are entered, and the Secretary shall certify that he has compared the sketch with the animal known by him to be the animal described, and that the pedigree correspond with the records in his office. He shall also brand on the hoof of the animal its Herd Register number. When thus completed, the certificates shall be closed and sealed by the Secretary of the Herd Book certifying to same, and sent by mail to the Secretary of the American Guernsey Cattle Club for identification and authentications of the cattle on arrival.

The name of the ship bringing the cattle, the port and date of landing shall be filled in after the certificates are opened by the Secretary, and they shall remain in the office of the Club as an evidence of importation.

7. Breeders shall furnish a certificate of service of dams and the name and Herd Book number of the bull, if served before shipping from the Island or England.

8. Dams of calves imported in dam, or dropped at sea, must be registered before their calves.

9. The names of the Island or English breeders of imported animals shall be printed in the American Herd Register, with the names and Island or English Herd Book numbers of their sires and dams, unless such sires and dams are already in this country and eligible for entry. Where it is found to be impossible to procure Island or English Herd Book names and numbers of sires and dams of imported animals, such animals can only be admitted to the registry in the American Herd Book by consent of the Executive Committee, or by vote of the Club.

10. The same name shall not be given to more than one animal without a prefix or suffix. Slight changes in spelling shall not constitute an original name, and the numerals 2, 3, 4, etc., shall only be allowed to actual progeny of the first animal bearing the name.

11. All Guernsey cattle imported for public sale must be registered in the American Guernsey Herd Book before such sale, or they cannot be registered except by special vote of the Executive Committee or by vote of the Club.

12. APPLICATIONS FOR HOME REGISTRY.—The Secretary shall furnish forms on which applications for entry shall be made, stating the sex, the name, the color, and marks as fully as possible; the name and address of breeder; the date of birth; names of all the owners; Herd Register number, with the exceptions noted in Section 9, and name and number of sire and dam; signature of the applicant, who, if not the breeder, must furnish the breeder's certificate. And in the outline sketches

on the back of each form, the white markings of the animal shall be drawn, and the colored parts shaded or indicated by letters, and certified by the applicant.

13. TRANSFERS.—Every change of ownership must be accompanied by a certificate of transfer. If these cannot be procured, transfers can only be made by vote of a majority of the Executive Committee, or by vote of the Club.

Certificates of transfer of animals in calf shall contain also certificates of the date of service, and name and number of bull.

14. It shall be the duty of each member to inform the Secretary of the date of the death of all registered animals, to be recorded by him.

15. No change shall be made in the By-Laws of the Club, unless due notice has been given of proposed change to all members in the call of the meeting at which it will be considered.

Quartz 4022.
Owned by Mr. Robert W. Lord, The Elms, Kennebunk, Me.

Rules and Instructions for Entry and Transfer of Cattle.

No animals shall be registered or transferred unless the following conditions are complied with:

They must be described and sketched on blank forms furnished by the Secretary.

They must be imported; or traceable through both sire and dam to animals imported from the Island of Guernsey.

Imported animals must be registered in one of the *Island* Herd Books, or in the Herd Register of the English Guernsey Cattle Society, and no imported animal is eligible for entry unless so registered.

All fees must be paid in advance. The Secretary is authorized by vote of the Club not to sign certificates until the fees are paid.

Make all remittances payable to *Wm. H. Caldwell, Treasurer.*

Remit by draft on Boston, New York, Philadelphia, or Chicago.

Personal checks on local banks should not be sent *unless they are drawn* on a bank in the above named cities. At local banks draft or cashier's check can be secured on banks in cities named.

Money can also be sent by postal order or by registered mail.

FEES (See also Article 3 of By-Laws).

Home-bred animals offered under six months of age,	$2.00
Home-bred animals offered over six months of age,	3.00
Imported animals offered within six months after landing,	2.00
Imported animals offered over six months after landing,	3.00
Members of the Club pay one dollar less.	
Transfer fees for each animal, to all,	$1.00

Firms or partnership cannot benefit by *members'* fees, unless each partner is a member of the Club.

The seller must pay all registry and transfer fees.

Every transfer of ownership is recorded. Ownership will not descend in closing estates without transfers.

The sketch and description must be carefully filled out, and color and distinguishing marks given.

The application should be filled out *with ink* and *signed upon both sides*.

Special attention is called to the following:

Every animal must be registered by its *breeder* or *importer*. The breeder is the owner of the dam at time of service. Should the animal be sold before the calf is dropped, the seller must certify to the service on the application for transfer. Should the dam with the calf be sold *after* the birth of the calf but before it has been registered, the calf will have to be registered by the owner of the dam at birth of the calf. Buyers should never buy animals which *can* be registered, but *insist* on registry before completing a purchase. Neglect of this makes infinite trouble to the buyer. Applicants should remember that all common and most single names are taken.

The same name shall not be given to more than one animal. Its offspring can be 2d, 3d, etc., but the offspring of 2d, 3d, etc., must change the name. A prefix or suffix can be added to perpetuate a favorite name, viz.: *Royal* Jeweller or Jeweller's *Jessie*.

Neglect of early entry of animals in the Register usually causes much delay and correspondence. The Club in the beginning recognized it a bad policy to delay registration. The fees were accordingly arranged to give a premium on early entry in the books; after an animal passes six months of age, the fee is $1 more. Some parties think this is an injustice, as it compels the entry before one can fully judge the capability of their cattle. This may be true to a certain extent, but as animals are often sold at an early age, they should not be allowed to go unregistered. Whenever they do, it presents many annoying complications. All breeders are urged to make early entry of their animals, and to preserve a copy of the sketch of the animal as sent to the office for registration. They will thus have always at command means of identification of their herd. This is especially important when changing help or in settling estates. The office has published a Private Herd Book in two forms to aid breeders in keeping track of their herds.

As the pedigree of an animal is becoming of more and more importance, the closest attention is given to its correctness. Inasmuch as all such records are dependent largely upon honor, every person proved to have made willful misstatements either in regard to the birth, or pedigree, or qualities of his animals, forfeits all consideration from the Club, and is wholly debarred from its records. Several such cases have come up and been dealt with very sharply and promptly. Sketches of all registered Guernseys are kept in the Secretary's office, and any buyers who are in doubt whether they have bought the right animal can satisfy themselves by comparing sketches of it with the original sketches.

When the service bull is owned by another person than the one owning the cow, a certificate of service must be obtained from the owner of the bull on blanks that are furnished upon request from the Club's office.

Tricksey 1760.

Property of Geo. C. Hill & Son, Rosendale, Wis.

"Aside from a light jaw, she was a model cow."—C. L. Hill, "The richest cow I ever had."—R. S. Huston.

Record 2 lbs. 7¾ oz. of butter in one day.

Tricksey was among the importations of the late S. C. Kent of West Grove, Pa. She was sold to Mr. I. J. Clapp of Kenosha, Wis., and finally owned by the Minn. Agl. Experiment Station. Mr. Kent wrote Mr. Clapp that all things considered, he thought Tricksey the best cow he ever owned.

She has left a number of very noted descendants, as the pag of this issue record. Besides the cows appearing in the lists of records, she is the dam of Mr. Francis Shaw's noted bull Tricksey Squire 2542.

CATTLE BREEDER'S PERPETUAL CALENDAR.

Day of Month Served	SERVED IN JANUARY WILL CALVE	SERVED IN FEBRUARY WILL CALVE	SERVED IN MARCH WILL CALVE	SERVED IN APRIL WILL CALVE	SERVED IN MAY WILL CALVE	SERVED IN JUNE WILL CALVE	SERVED IN JULY WILL CALVE	SERVED IN AUGUST WILL CALVE	SERVED IN SEPTEMBER WILL CALVE	SERVED IN OCTOBER WILL CALVE	SERVED IN NOVEMBER WILL CALVE	SERVED IN DECEMBER WILL CALVE
1	Oct 13	Nov 13	Dec 11	Jan 11	Feb 10	Mar 13	Apr 12	May 13	June 13	July 13	Aug 13	Sept 12
2	14	14	12	12	11	14	13	14	14	14	14	13
3	15	15	13	13	12	15	14	15	15	15	15	14
4	16	16	14	14	13	16	15	16	16	16	16	15
5	17	17	15	15	14	17	16	17	17	17	17	16
6	18	18	16	16	15	18	17	18	18	18	18	17
7	19	19	17	17	16	19	18	19	19	19	19	18
8	20	20	18	18	17	20	19	20	20	20	20	19
9	21	21	19	19	18	21	20	21	21	21	21	20
10	22	22	20	20	19	22	21	22	22	22	22	21
11	23	23	21	21	20	23	22	23	23	23	23	22
12	24	24	22	22	21	24	23	24	24	24	24	23
13	25	25	23	23	22	25	24	25	25	25	25	24
14	26	26	24	24	23	26	25	26	26	26	26	25
15	27	27	25	25	24	27	26	27	27	27	27	26
16	28	28	26	26	25	28	27	28	28	28	28	27
17	29	29	27	27	26	29	28	29	29	29	29	28
18	30	30	28	28	27	30	29	30	30	30	30	29
19	31	Dec 1	29	29	28	31	30	31	July 1	31	31	30
20	Nov 1	2	30	30	Mar 1	Apr 1	May 1	June 1	2	Aug 1	Sept 1	Oct 1
21	2	3	31	31	2	2	2	2	3	2	2	2
22	3	4	Jan 1	Feb 1	3	3	3	3	4	3	3	3
23	4	5	2	2	4	4	4	4	5	4	4	4
24	5	6	3	3	5	5	5	5	6	5	5	5
25	6	7	4	4	6	6	6	6	7	6	6	6
26	7	8	5	5	7	7	7	7	8	7	7	7
27	8	9	6	6	8	8	8	8	9	8	8	8
28	9	10	7	7	9	9	9	9	10	9	9	9
29	10		8	8	10	10	10	10	11	10	10	10
30	11		9	9	11	11	11	11	12	11	11	11
31	12		10		12		12	12		12		12

Period of gestation calculated at 285 days.

Left hand column, date of service; column for month, date of calving.

Imp. Princess May XI. 10798, A. G. C. C.
(R. G. A. S. 3108 P. S.)
Imported in 1898 by J. N. Greenshields, Isaleigh Grange Farm, Danville, P. Q.